THE EVERYDAY BLACKSMITH

WITH MULTIPLE VARIATIONS FOR STYLES AND FINISHES

LEARN TO FORGE 55 SIMPLE PROJECTS YOU'LL USE EVERY DAY

NICHOLAS WICKS

Blacksmithing can be a dangerous activity. Failure to follow safety procedures may result in serious injury or death. This book provides useful instruction, but we cannot anticipate all of your working conditions or the characteristics of your materials and tools. For your safety, you should use caution, care, and good judgment when following the procedures described in this book. Consider your own skill level and the instructions and safety precautions associated with the various tools and materials shown. The publisher cannot assume responsibility for any damage to property or injury to persons as a result of misuse of the information provided.

Quarto.com

© 2019 Quarto Publishing Group USA Inc.
Text © 2019 Nicholas Wicks
Photography © 2019 Quarto Publishing Group USA Inc.

First Published in 2019 by Quarry Books, an imprint of The Quarto Group,
100 Cummings Center, Suite 265-D,
Beverly, MA 01915, USA.
T (978) 282-9590 F (978) 283-2742

All rights reserved. No part of this book may be reproduced in any form without written permission of the copyright owners. All images in this book have been reproduced with the knowledge and prior consent of the artists concerned, and no responsibility is accepted by producer, publisher, or printer for any infringement of copyright or otherwise, arising from the contents of this publication.

Every effort has been made to ensure that credits accurately comply with information supplied. We apologize for any inaccuracies that may have occurred and will resolve inaccurate or missing information in a subsequent reprinting of the book.

Quarry Books titles are also available at discount for retail, wholesale, promotional, and bulk purchase. For details, contact the Special Sales Manager by email at specialsales@quarto.com or by mail at The Quarto Group, Attn: Special Sales Manager, 100 Cummings Center, Suite 265-D, Beverly, MA 01915, USA.

11

ISBN: 978-1-63159-712-1

Digital edition published in 2019
eISBN: 978-1-63159-713-8

Library of Congress Cataloging-in-Publication Data
Wicks, Nicholas, author.
The everyday blacksmith : learn to forge 55 simple projects you'll use every day, with multiple variations for styles and finishes / Nicholas Wicks.
ISBN 9781631597138 (ebook) | ISBN 9781631597121 (pbk. : alk. paper)
1. Blacksmithing--Amateurs' manuals.
LCC TT221 (ebook) | LCC TT221 .W53 2019 (print)
DDC 682--dc23
LCCN 2019011816 (print) | LCCN 2019013562 (ebook)

Design: Burge Agency
Photography: Matt Kiedaisch/Outsider Media
Illustration: Mark Coletti/Moose Art Designs

Printed in China

To Edward A. Wicks
Metalworker
1923–2018

Thanks, Papa, for passing on your
love of sharing stories. I think you
would have liked this one.

CONTENTS

INTRODUCTION

PART I:

TOOLS AND TECHNIQUES

PART II:

PROJECTS

CONCLUSION

APPENDIX

This 100-year-old Champion coal forge was marketed to home smiths and farmers who needed to make and repair everyday items such as plowshares (the bade of a plow). Today, there is a resurgence of home smiths restoring these old tools and using them to make the next generation of everyday blacksmith goods.

→

INTRODUCTION

BLACKSMITHING TODAY

A village blacksmith from the 1800s walking into today's blacksmith shop would find many familiar sights. The anvil, forge, hammers, tongs, and vise in the modern shop may even have been made during his time. He would see a kindred spirit in his modern counterpart, hunched over the anvil bringing a piece of hot iron to life with rhythmic blows of the hammer. What our village smith would not recognize would be the power tools—welders, drills, grinders, and power hammers that save the modern smith both time and sweat. Most of all, he might be surprised by the products of our modern smith and the blacksmith's evolving role in his—or her!—community.

For centuries, blacksmiths were the craftsmen and artists who worked society's most important material—iron. The village blacksmith was the local hardware store, contractor, and mechanic all in one—recall, this was during the time of horses. Blacksmiths were not only a fixture in their community, they helped shape that community through their particular method of making the hinges, hooks, brackets, and tools their neighbors used every day.

In a twist of irony, the age of mass production that the blacksmith helped usher in almost led to the blacksmith's demise. Yet, instead of going the way of the dodo, blacksmithing today is enjoying a resurgence. Our modern society discovered that no amount of technical perfection replaced the feeling of picking up a hand-forged object and knowing someone's creativity and effort went into shaping that particular piece. In makerspaces, art schools, blacksmith associations, modern shops, traditional smithies, online forums, and makeshift home forges, like-minded tinkerers have not only revived the craft, they are taking it in new and exciting directions.

Where do your interests lay? Traditionalists enjoy the history and culture of the craft and work mostly without modern power tools or electric welders. Artists and sculptors, drawn to the versatility and permanence of steel, explore the design limits of the material. Professional smiths are tasked with balancing an artistic bent within the architectural, functional, and financial constraints of commissioned projects and often combine traditional and modern techniques. Home smiths usually fall somewhere in between. They make pieces to use or sell while also dealing with constraints of time, space, funding, and equipment.

Many of these categories overlap. For example, in my limited years blacksmithing, I've been a hobbyist in a home forge, an apprentice in a professional shop, a part-time blacksmith working out of a friend's barn, and finally, owner of my own shop. In that time, I have been asked to make cannons and Samurai swords (jobs I sadly turned down) as well as traditional catapults and Nordic runes (jobs I couldn't refuse). I've restored axes, designed custom gates, rebuilt truck beds, forged hand tools, furnished homes with hardware, created garden sculptures, and welded broken tractors parts. Many of these requests were things I didn't know could even be needed until a neighbor would call up and ask the always enchanting question that starts with "can you make…?"

The beauty of blacksmithing is that answer, more often than not, is yes. For the world of steel is our domain. So let us return to our 1800s village blacksmith and ask this question: Is the part we play in our community today so different? We may not be repairing hammers or making nails for the local carpenter (actually, see the hardware chapter), but we blacksmiths continue to serve our community with the products of our craft. Honoring that tradition, I hope this book helps you—like our blacksmithing forebears—continue to shape our world with unique pieces for yourselves, friends, neighbors, and community to use and enjoy every day.

BOOK STRUCTURE

This book is divided into two parts. Part I is a reference of shop basics: safety, equipment, and techniques. Basic techniques are illustrated through a series of projects. It also covers methods for finishing pieces.

Part II, the majority of this book, provides a step-by-step guide to a diverse range of blacksmith projects. Projects are structured by category and difficulty. The emphasis of this book is on the accessibility of techniques, functionality of projects, and diversity of design. The projects selected for this book were ones that can be made, used, and enjoyed by folks in a variety of settings. With care, someone with only basic tools and equipment will be able to make every piece in this text. New techniques are explained the first time they appear in a project and are indexed for reference in the back of the book. There are many approaches to projects, influenced by available time, materials, tools, techniques, and one's aesthetics, and these pages represent only one of many ways to do things. Treat project steps as suggestions rather than scripture, and find out what works best for you.

I've tried to account for the great diversity in our craft today by featuring projects by a range of smiths from several countries, up and coming and established, traditional and modern. Their stories are often as inspiring as their work. Look them up. As you will soon see, there are many exciting projects waiting for you to explore in the following pages. For newer smiths, I hope this book both teaches and motivates you as you continue on your journey in blacksmithing. For established smiths, I hope this book provides ideas, a launching pad, and timesaving when designing your own new products.

What are you waiting for? Get smithing!

PART I:

TOOLS AND TECHNIQUES

CHAPTER 1:

SETTING UP SHOP

This section covers basic blacksmithing concepts. The emphasis is on helping you get started making the pieces in this book quickly. If you are brand new to smithing, it will be very helpful if you supplement this text with an additional reference on blacksmithing basics or take an introductory blacksmithing class. And if you are not already a member, consider joining a local or national blacksmithing association. These associations exist primarily to teach and help new smiths get into the craft. Many blacksmiths may be big and scruffy, but they are also friendly. We all learned our trade from generations before us, and most are happy to share with the next generation.

When setting up your shop, also known as a forge or smithy, there are certain immediate choices to make. This includes what type of fuel source to use and what basic tools to acquire or make. The pros and cons of various options are discussed as well as the range of tools useful for making all of the projects in this book. Blacksmiths are diverse. Depending on your interests, location, and budget, different setups may be right for you. You may not always know what you prefer until you try something out—that's part of the adventure!

BLACKSMITHING. WHAT'S IN A NAME?

The two most common questions I get when I tell people I am a blacksmith are "Do you shoe horses?" and "Can you make knives?" For the record, although there is often crossover between each trade, farriers specialize in shoeing horses and bladesmiths make knives.

Today's blacksmiths are quite varied. Some specialize in making tools, others in public sculptures. Some work only with traditional tools and materials, others use modern technology. Many do a bit of everything. Regardless, all blacksmiths are related by the type of material they work (steel nowadays) and the way in which they work it (forging). When steel is heated, it becomes more malleable, allowing it to be shaped by hammering and other means. This process is called forging. While today's blacksmith may use modern welders and may work in materials in addition to steel, heating and working steel remains core to the craft.

SAFETY

A few years ago, I took a motorcycle safety course. The instructors called motorcycling serious fun—fun because it was enjoyable, serious because you can get seriously hurt without proper safety. The same can be said for blacksmithing. Few activities offer more potential for injury to yourself, others, and property than blacksmithing. Take precautions seriously and it will allow you to enjoy blacksmithing fully.

A leather glove on the non-hammer hand and a weightlifting glove on the hammer hand can balance protection and dexterity. ←

SHOP

Excited to finally get your shop together? Let's keep it around for a while by not burning it to the ground. Keep multiple extinguishers about and always be aware of the relationship between flammable materials, the forge, and the hot steel you are working.

EYES

Even though you have two eyes, you don't have any to spare. A good practice is to always wear safety glasses anytime you are in the shop, regardless of activity. A full-face shield is also useful for heavy grinding.

EARS

Blacksmithing is extremely noisy. Ear protection is thus a great idea. I prefer earplugs. That way, if I have to switch between additional gear like a respirator or welding helmet, I don't have to worry fitting earmuffs on my head as well.

HANDS

The non-hammer hand handles the steel from the forge and is most exposed to heat. Inexpensive leather gloves are a good option for this hand as the heat from the forge destroys gloves fast by burning through the stitching and shrinking the leather (you can tell a blacksmith shop because it will often have a pristine pile of gloves for one hand and a complete shortage for the other hand).

LUNGS

Fumes from your forge, whether coal or propane, should be minimized with good ventilation. A good respirator with filters (such as 3M 2097 P100) is a must if you plan to engage in cutting, grinding, and welding with power tools. Those activities produce extremely harmful fumes and particulates that go right through typical dust masks.

CLOTHES

Use good boots as well as duck canvas pants or similar. Always tuck boot cuffs inside your pant legs so a hot piece of metal can't find its way inside your boot. On your upper body, long sleeve work shirts will help reduce burns from sparks and grinding. Many blacksmiths also wear leather aprons. Leather not only protects better than flannel, it completes the blacksmith "look."

BASIC TOOLS

The following are the essential tools to carry out the basic functions of blacksmithing and make the projects in this book. It is not a comprehensive list. A few others are discussed that will greatly increase your ability to work faster. Be warned, "blacksmith" is actually another term for a compulsive tool hoarder. Many blacksmiths have multiple tools bought on impulse that are now gathering equal parts dust and shame in some dark corner of their shop. If you ever find yourself trying irrationally to justify that next tool, just remember that 200 years ago blacksmiths with no electricity and far fewer resources than you or I seemed to get by just fine.

The essential tools of the blacksmith haven't changed much over the years. This propane forge (centered above) is the only tool here made since the end of World War II.
↑

FORGE

The forge heats metal to the point where it can be worked by hand. The two main types of forges are defined by their fuel source.

COAL FORGES

Coal forges use coal or coke (coal that has been heated to remove organic material) to heat steel. They are simple to build, even rudimentary ones can achieve forge-welding heat, and they can accommodate many different-shaped pieces. The disadvantages of coal forges are they require dedicated ventilation, if indoors, and constant management of the fire.

GAS FORGES

Unlike coal forges, which are open, gas forges heat an enclosed space in which the metal is placed, usually with propane. Gas forges are simple to operate and can heat multiple pieces at the same time without burning. However, they are more complex to set up initially, the size pieces that can be worked are limited by the size of the forge enclosure, and it is harder to achieve welding temperatures in lower quality gas forges. Either type can be built or purchased, but gas forges are more complex to build.

ANVIL

The anvil is the primary surface with which steel is worked. When one hits down with a hammer, the anvil acts as a surface to "push back" against the blow, allowing the metal to be moved and shaped. The anvil is one of the greatest initial expenses for the smith. If purchasing new, avoid anvils made completely of cast iron. They will deform quickly. If purchasing used, look for cracks in the body of the anvil (avoid those!). A slightly chipped face can be cleaned with grinding. If building an anvil, railroad tracks are a common starting material. The preferred height of the anvil for minimal back strain is at knuckle height while standing. The base can be made of metal or wood.

This London-pattern anvil is one of the most common types found today. ↓

TYPES OF STEEL

While you don't need to be a chemist, there are a few key terms with which the blacksmith must be familiar. The material with which today's blacksmith works most is "mild steel," short for mild carbon steel. All steels are an alloy (a metal composed of multiple elements) made up primarily of the elements iron and carbon. Mild steel has a low carbon content and a nice balance of strength and malleability. It is similar in composition to "wrought iron," the original material of the blacksmith. Although still used in specialty situations, wrought iron is no longer produced in large quantities. There are also "high carbon," or tool steels. The higher carbon content of these steels allows them to be "heat treated," a process that enable tools to hold their shape under stress (see the Tools chapter in Part II). "Stock" is a term used to identify starting materials, such as "round stock" or "flat stock." Steel can be purchased online or from local metal suppliers.

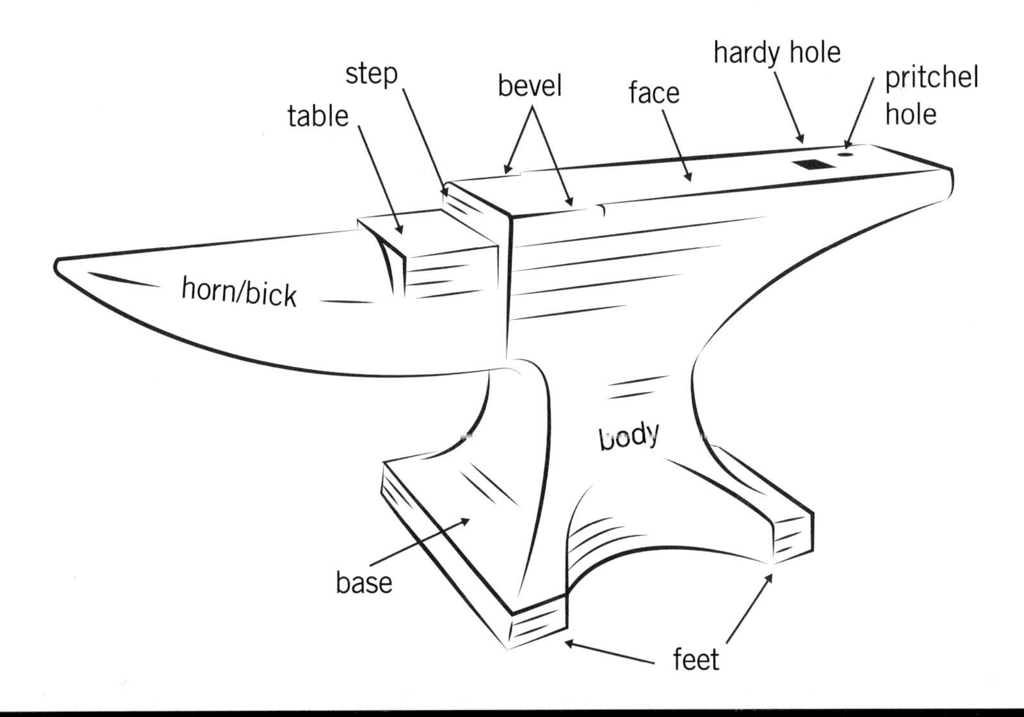

The blacksmith makes the majority of his or her pieces with a small assortment of hand tools.

VISE

Several operations in blacksmithing require the steel to be held in place with a vise. Blacksmiths traditionally use a leg vise, which can be secured in the floor or to a stand. However, a sturdy bench vise will also suffice for starting out.

QUENCHING BUCKET

A quenching bucket is used to cool material being worked as well as tongs and other hand tools. It should be a big bucket made of steel or wood so it doesn't melt when in contact with hot steel.

CUT-OFF TOOL

This tool, also known as a hardy chisel, fits in the hardy hole of the anvil and is used to cut steel. They can be purchased or forged from tool steel but will likely require a helper (known as a "striker") or power hammer because of the size starting material to move around. If you have access to a welder, one can be made from a splitting maul head that is cut down and welded to a piece of square stock.

HAMMERS

The flat end of a hammer is called the "face." It should be slightly rounded at the edges and slightly crowned (dome shaped) so that slight variations in the angle of hammer blows do not dig into the steel. The face can be rounded with a grinder. The end opposite the face is called the "peen." Ball-peen hammers have a ball-shaped peen and are used for dishing (that is, curving) shapes, texturing steel, and riveting. Cross-peen hammers have a peen that looks like a roofline of a house. They are used for fullering (spreading) steel. The blacksmith requires multiple sizes and types of hammers: at the minimum, a small ball-peen hammer in the range of 10 ounces (or just over ½ pound; 284 g) for fine work; a medium ball-peen hammer and a medium cross-peen hammer in the 1½ to 2½ pound (680 to 1134 g) range for general work; and a large hammer—either ball peen or cross peen—in the 3-plus pound (1.4 kg) range for heavy work. While not a beginning project, hammers can be made with only a few specialized tools.

TONGS

Tongs are used to hold and handle hot steel. It is extremely important that tongs fit the steel being worked. A loose piece of hot metal is tough to work and can cause a lot of damage to you and others. There are a few multiuse tongs that together can accommodate several jobs. Making a good pair of general-use tongs is covered in the Tools chapter.

CHISELS, PUNCHES, AND DRIFTS

Chisels are used for stamping and cutting steel. Punches and drifts are used to make and then enlarge holes in the steel. Chisels and punches should be made from tool steel. Drifts can be made from mild steel. Making each is covered in the Tools chapter.

FORKS AND WRENCHES

Bending forks, or scrolling forks, are used to make precise bends in steel. Twisting wrenches are used to create twists along the length of a piece of stock. Each can be made or purchased. Making each is covered in the Tools chapter.

WIRE BRUSH

As steel is heated and worked, the surface oxidizes (mixes with oxygen) and creates a "scale" on the piece. Wire brushes are used to remove this scale during the forging process and after a piece is finished. Wire brushing is discussed in the next chapter.

FILES

Files are used to remove material and smooth surfaces. They come in various cuts (how fine the teeth are) and shapes. It is useful to have a rasp for heavy filing as well as a half-round file and a round second cut (less coarse) file.

DRILL

Although holes can be made by punching, drilling allows for holes to be made with great precision while the material is cold. I highly recommend a drill press in addition to a hand drill.

ADDITIONAL TOOLS WORTH EVERY PENNY

ANGLE GRINDER

The most versatile tool for working metal, it can be used to cut, grind, and clean steel. A 4½" (117 mm) grinder without a paddle switch is my preference, so that it can be operated in multiple hand positions. Useful attachments are cut-off wheels (cutting), grinding wheels (removing large amounts of material), flap discs (removing precise amounts of material), and wire brush caps (removing scale).

WELDER

Perhaps the ultimate time-saving machine, an electric welder will allow you to permanently join metals with the pull of a trigger. MIG welders are easy to use and great for general work. Buy the best quality your budget will allow, but even a cheap welder will quickly become indispensable.

OXY-ACETYLENE TORCH

An oxy-acetylene torch combines oxygen and acetylene or propane into a high-temperature torch flame. They can be used for cutting steel, making precision bends (by only heating specific parts of the steel), making rivets, and even for welding.

MAKING VERSUS BUYING TOOLS

Traditionally, blacksmiths made many of their own tools out of necessity. Even today, unique jobs require specialty tools or jigs the blacksmith has no choice but to make. However, as modern smiths, we largely have the luxury to choose what we purchase and what we make. If making tools interests you, then pursue it. If you have access to ready-made tools and your interests lie elsewhere, that's fine too. Starting out, some necessary basic tools such as a hardy cut-off require some advanced techniques and the use of tool steels, so purchasing might not be a bad idea.

THE PROCESS OF FORGING

The basic actions of the blacksmith involve heating, drawing out (elongating), spreading, bending, twisting, cutting, punching, upsetting (compressing), and cleaning steel. The blacksmith uses the hammer in several specialized motions to carry out these actions, which are diagrammed here and referenced throughout this book.

HAMMER ACTIONS

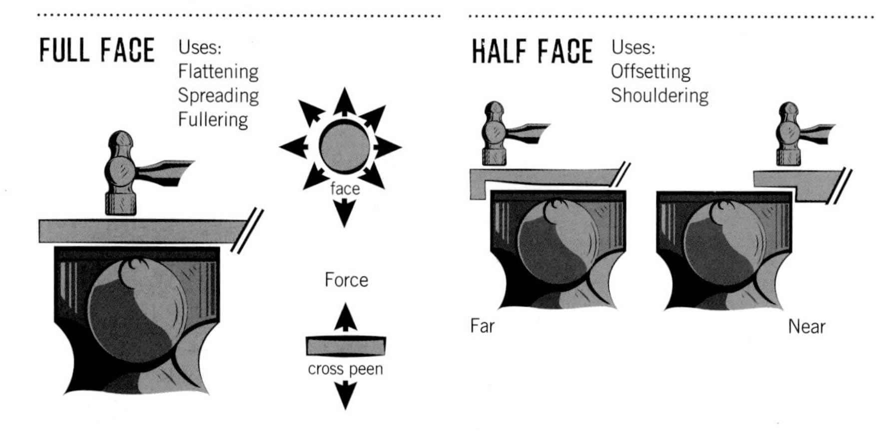

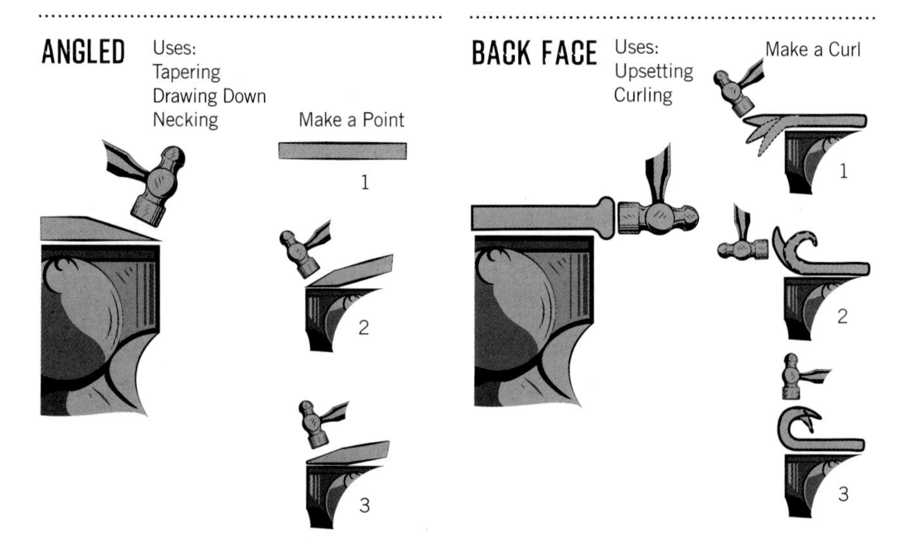

LEARNING A NEW CRAFT

Some people experience difficulty with blacksmithing for three main reasons, and they are unrelated to skill or strength. The main reason is a lack of understanding of blacksmithing terminology. As in any specialty, blacksmithing has its own jargon that must be learned. Some of the language is intuitive—like punching holes—and some of it is confusing at first. For example, it is impossible to execute a half-face blow if one doesn't know what a half-face blow is (a hammer blow that is half on the face of the anvil and half off the anvil). To help learn this new language, specialized terms are defined the first time they appear in the text and there is also a glossary in the back of the book.

The next reason someone may have trouble is because the actual tools, techniques, or steps being described are unfamiliar. If you are unsure about a tool or process—like how a fullering tool (a forming tool for spreading material) operates, look it up in this book or search online to clarify.

Finally, it is important to progress at the proper gradient, becoming proficient in basic techniques before moving onto more complex actions. Projects in each chapter are organized by difficulty. If you find yourself having a great deal of trouble with a certain project or technique, do a simpler one first.

FORGING ACTIONS

DRAWING

Elongating

Tapering

Spreading

BENDING

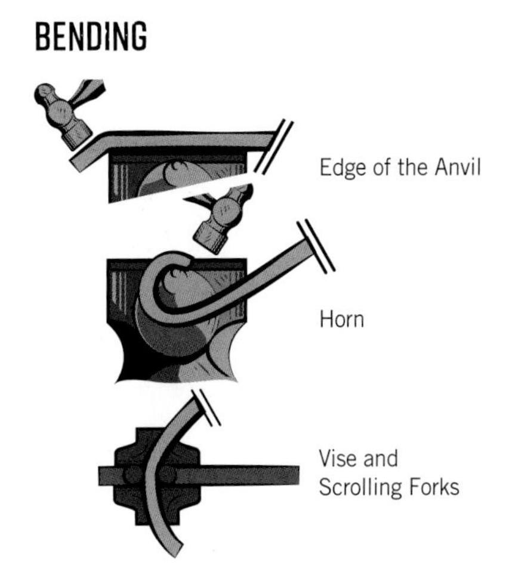

UPSETTING

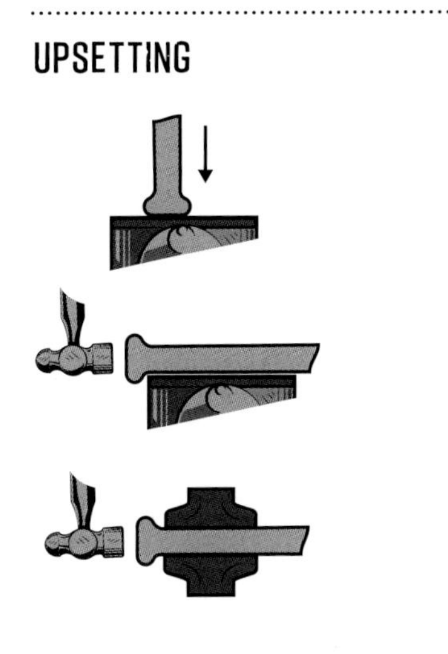

PUNCHING

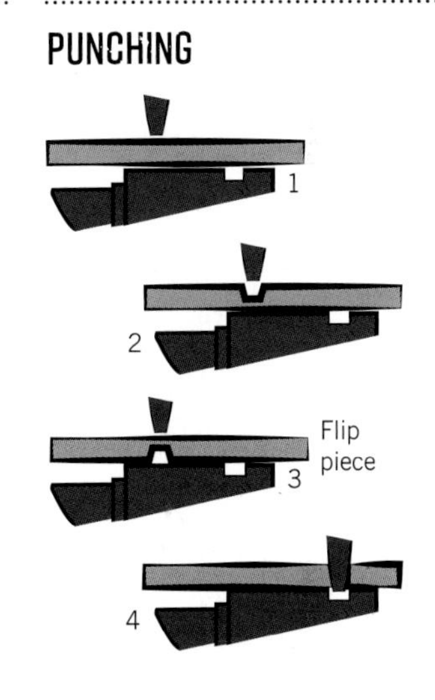

CUTTING

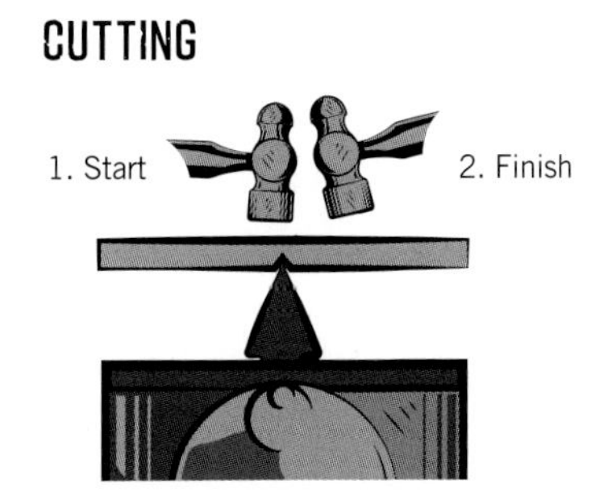

WELDING

CHAPTER 2:

FINISHING PIECES

Once a piece is forged, the last step to finish metalwork is the process (aptly named) of finishing. Unprotected, mild steel will rust. The first consideration is, thus, how to best protect from corrosion. The other considerations are use and aesthetics. Is the piece going to be indoors or outdoors? A light wax sufficient to protect an indoor hook for years wouldn't last one rainfall on an outdoor hanger. Do you want a natural finish or bright colors? Will the object be in contact with food? These questions can help determine your best choice. Don't be overwhelmed by the numerous finishing methods one can find online. The best smiths I know work with a limited number of simple finishes but do so expertly. Experiment until you find combinations that work for you.

Finishes from left to right: piece before scale removal, brushed and waxed, brass brushed, linseed oil, matte black paint, rusted.
←

REMOVING SCALE

All finishes require the removal of scale from the finished metalwork to varying degrees. This is the first step in the finishing process. Here are some options.

Hand brushing: Suitable for moderate burnishing, best done when the piece is still hot.

Wire wheel for a bench grinder: Good for greater scale removal on smaller pieces.

Wire brush for an angle grinder: Good for fast cleaning of larger pieces. Wire brushing with a bench or angle grinder is extremely hazardous, so follow proper operating procedures.

Wire wheels or cups for a drill: Useful if you only have access to a drill, slower but good for getting into nooks.

Sandblasting: Among the most abrasive methods, best for situations that demand maximum scale removal.

Tumblers: An amazing time-saving tool for large quantities of smaller items, but you lose some of the forged surface textures.

Vinegar: Helps loosen scale prior to burnishing if pieces are soaked overnight.

WAX

The process of heating and hand working steel create a beautiful surface texture unique to our trade. A wax finish highlights and preserves this surface. It is my preferred choice for interior pieces. After wire brushing the completed piece to a sheen, I use a product called "Renaissance Wax." It is a combination wax and polish that brings out surface variations beautifully. Use a brush to apply, wipe away the excess, and then buff the surface lightly with a cloth. It can be applied cold. Apply two or more coats several hours apart. It is not food safe or suitable for outdoors. A traditional food-safe option is to heat the steel to a black heat and melt beeswax into the surface. Wipe away the excess and lightly buff.

OILS

Oil finishes create a darker look, usually in the blue to black range. Have you ever cooked with a cast iron pan? If so, you should be familiar with the ritual of seasoning the pan. Putting an oil finish on your metalwork is a similar process. Linseed oil is a good food-safe choice. Heat the metal to a black heat. Wire brush the finished piece to remove excess scale. If it is a small piece, quench directly in a bucket of the oil. With larger pieces, use a brush or rag to apply to a section of the piece at a time. If the oil ignites when in contact with the metal, it is too hot. Repeat at a lower temperature. Wipe the excess oil off with a cloth. Because oil seeps into the surface of the steel, it is more resistant to moisture than wax finishes, but still not suitable for permanent outdoor pieces without additional sealing such as lacquer.

PAINTING BLACK

Many blacksmiths loath having to paint a custom piece black. What's with all the black gates then? While it may be a traditional choice, paint in general, and black specifically, mutes the fine detail of hand forged pieces. This is the same reason machine-made pieces are always painted (to hide ugly surfaces and welds created through fabrication). Where painting is necessary, try satins and colors like gray that help highlight rather than hide the details of your work.

PAINTS AND LACQUERS

If you want to preserve the waxed or oiled look for pieces exposed to moisture (such as outdoor pieces or bathroom fixtures) cover the finished piece with a good lacquer. Although clear, they will darken and gloss the existing surface. Painting is an inexpensive and simple option where good moisture protection or color variation is desired. Remove scale before applying primer and paint as loose scale can flake off over time, taking the painted surface with it.

PATINAS AND STAINS

Patinas and stains are a nice way to get rich colors without losing interesting surface variation. Individual procedures vary, but key aspects are as follows: meticulous surface preparation; careful application, and ample surface protection afterward. Sandblast pieces or wire brush heavily to remove all scale. Patinas generally look bad on oxidized steel (scale). Clean the piece with alcohol and afterwards handle with latex gloves. Even fingerprints will disrupt patinas and stains. Follow the application procedures for the particular patina or stain and seal with a high-quality lacquer. Sculpt Nouveau has a wide selection of products as well as helpful technical support. Combined with a high-quality lacquer, they are suitable for outdoor pieces.

MIXING AND MATCHING

Black magic. Many blacksmiths use a variation of this cocktail: equal parts beeswax, turpentine, and linseed oil, and a small amount of Japan drier. Melt together the mixture using an electric (no-flame) heat source. Once cooled, apply like wax. Turpentine allows for better penetration into the steel, while the Japan drier speeds the drying process.

Brass accents. While the piece is at a dull red heat, burnish with a brass wire brush. Some of the brass will impregnate the steel, giving a brassy tint.

Painted graphite. If you must use black paint, mix in a very small amount of graphite and lightly burnish the surface with a cloth just before the paint is dry to the touch. This creates a black finish with more textural interest.

Rusted. This process can be induced with a mixture of 10 parts hydrogen peroxide to 1 part vinegar and salt. Apply with a spray bottle. Remove excess and apply a lacquer sealer when desired effect is achieved.

PROJECTS

- Can Tab Opener
- Bookmark
- Spoon
- Barbecue Fork
- Leaf
- Skull
- Hook
- Calipers
- Bottle Opener
- Russian Rose

CHAPTER 3:
10 PROJECTS TO START

These starting projects highlight some of the basic techniques referenced throughout the rest of this book. They also provide a good gradient for practicing fundamental hammer skills. The importance of having this foundation of technique cannot be overstated. As you will see in these projects, even the simplest pieces require the smith to execute multiple techniques smoothly and quickly. Steel becomes harder as it loses its heat, making it more resistant to manipulation. Accordingly, the faster one is able to work, the more malleable the steel. Developing basic skills is thus doubly challenging for the blacksmith because of the need to "strike while the iron is hot" while still figuring out the proper way to strike.

With this in mind, there are really two rounds of skills to be developed from these introductory projects: the ability to make the projects themselves and then the efficiency to make them faster. Once you have a grasp of a given project, see if you can reduce the amount of "heats" (times in and out of the forge) it takes you to make it again while maintaining proper technique and form.

CAN TAB OPENER

SUGGESTED MATERIALS

3/8" (10 mm) round stock, run long (not cut to length yet)

SUGGESTED TOOLS

Small flat stock tongs

TECHNIQUES

Flattening round stock

Tapering

Curling

Can tab openers make for great promotional pieces since they are quick to make and easy to personalize. Stamp them with your maker's mark or customize them with a brewery's logo. Bottle openers are a more common blacksmithing product these days, but many craft breweries use only cans so it is good to offer both types of openers.

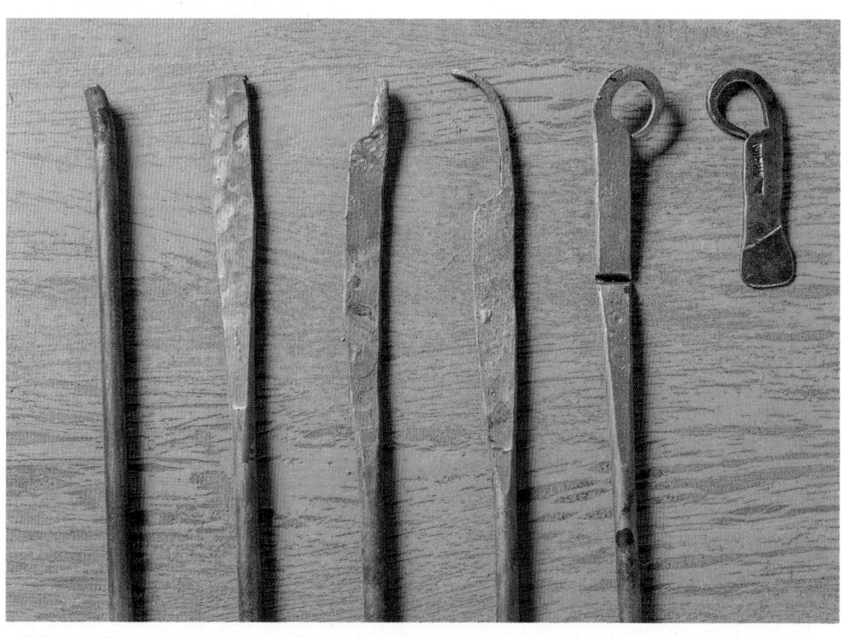

1. Heat up and flatten approximately 8" (200 mm) of round stock until it is ¾" (20 mm) wide.
→

2. With the narrow width against the edge of the anvil, use half-face blows to offset 1¼" (30 mm) of the flattened bar. Taper this offset to 2½" (65 mm).

3. Create a loop by curving the taper back into the flattened bar. Start the curve using angled blows on the horn of the anvil and then close it by flipping the piece over on the anvil and hitting the partially closed loop back toward yourself.
→

4. Cut off at 3½" (90 mm) and use small flat stock tongs to return the piece to the forge. Hold the cut-off side of the piece flat on the anvil at a 45-degree angle across your body (whether you are a righty or lefty, this will end up creating the tab opener at the correct angle) with the closer edge of the piece ¼" (6 mm) past the lip onto the anvil. Create the tab opener lip by hammering with half-face blows.

5. Clean the lip with a file or grinder. Wire brush and apply a finish to the piece

NEW TECHNIQUE: FLATTENING ROUND STOCK

Flattening round stock creates flat stock with rounded edges that complement the handmade aesthetic nicely (regular flat stock has sharper, mechanical edges). Flatten round stock using full-face blows, paying extra attention to hitting the metal head on to prevent gouging from your hammer. If you want the flattened steel to have a uniform final width, work back and forth along the whole length being flattened. If you want a more varied dimension, flatten one section at a time.

BOOKMARK

BY BOB MENARD, BALL AND CHAIN FORGE
PORTLAND, MAINE, USA

SUGGESTED MATERIALS

3/16" (5 mm) round stock cut to 6" (150 mm)

SUGGESTED TOOLS

Small round stock tongs

TECHNIQUES

Tapering

Flattening

Curling

The bookmark is a great project for beginning smiths. Within the project, many important early skills are learned. There are the three stages of tapering: square, octagonal, round, as well as producing a delicate curl with the hammer. A lot of basic hammer control forging the end flat is included, and the stock is lightweight and not too intimidating.

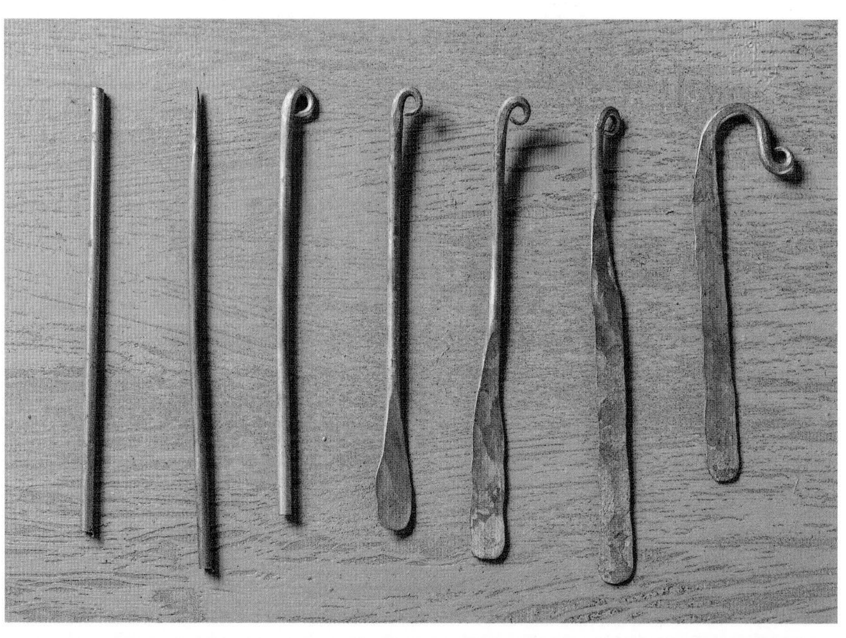

1. Taper the end to a point approximately 1" (25 mm) long and create a small curl on the pointed end.

2. Heat the opposite end of the piece and flatten back to within 1¼" (30 mm) from the starting curl, in the neighborhood of 1/16" (1.5 mm) thick and tapering it toward the end for a smooth transition. Working one section at a time creates some nice bulges that supplement a hand-forged look.

3. Form the larger curl using the same technique above but on the horn of the anvil, or use a ¾" (20 mm) pipe or similar as a mandrel (forming tool, see page 146). Don't extend the curl much past 180 degrees or close the space too much or it won't fit into the book easily.

4. Wire brush and finish.

NEW TECHNIQUE: CURLING

Creating curls pleasing to the eye involves working material in a deliberate and smooth fashion. Begin by holding the tip off the end of the anvil and hitting off the face off the anvil with light, angled blows. Work the tip slightly farther off after each hit, until a "C" shape is made. Flip the "C" over and begin to close the curl onto itself with backward and downward blows. The key to smooth curls is working the material evenly and not hitting too much in any given spot.

IN THE SHOP WITH BALL AND CHAIN FORGE

Whether he is it hosting classes out of his workshop in Portland or running educational programs through the Artist Blacksmith Association of North America (ABANA), Bob Menard has been teaching new smiths the art of blacksmithing for decades. He starts all new smiths with this project to get them practicing fundamental hammer skills. Try his favorite finish for these bookmarks, beeswax. It adds a nice warmth to the piece when handled.

SPOON

SUGGESTED MATERIALS

3/16" x 1" (5 mm x 25 mm) flat stock run long

SUGGESTED TOOLS

Small flat and round tongs

Fullering tool

TECHNIQUES

Fullering

Spreading

Drawing out stock

Great as a demonstration piece, spoons show how material can be manipulated in diverse ways through the forging process. It is important that the final cup of the spoon be smooth and completely free of scale so there are no crevices in which food can become lodged or bacteria can build up. Treat the spoon like a cast iron pan, oiling lightly after each use.

NEW TECHNIQUE: SPREADING

Use the fuller of the hammer to spread out the cup of the spoon. Begin in the center and work each side equally to create a symmetrical spread. After spreading the material, use the face of the hammer to smooth out the fullering marks.

1. Use a fullering tool or angled blows at the far edge of the anvil to isolate 1½" (40 mm) of the starting material. Forge a 3" (75 mm) taper leading to the isolated section and a slight taper on the end of the stock.

2. Spread the isolated tip into a spear shape.
→

3. Continue drawing out the handle to the desired final length. The spoon in this project is a cooking spoon with an 11" (280 mm) handle. Draw out the handle so that the top side of the end of the handle is about three times the width. Cut off and dress (clean) the end with a file or grinder.

4. Isolate 1" (25 mm) of the handle using half-face blows. Taper to 4" (102 mm). Bend the taper about 45 degrees away from the handle, make a very small curl in the same direction, quench the curl, then close loop using the horn of the anvil.

5. Reheat the spoon face and dish to form into the final shape.
↓

NEW TECHNIQUE: DISHING

There are a few methods for forming if you do not have access to a spoon former on a swage block. The simplest method is to dish with a ball-peen hammer using a former ground out of a stump. Another method is to use the head of a rounded railroad spike rounded and ground smooth. Use the face of the hammer to round the spoon over the spike's head. The latter method creates less hammer marks in the cup of the spoon.

6. Reheat the entire piece and curve the handle lightly into the final spoon shape.

7. Sand the spoon face to a smooth finish and coat the entire piece in oil.

BARBECUE FORK

SUGGESTED MATERIALS

5/16" x 1" (8 x 25 mm) round bar cut at 18" (457 mm)

SUGGESTED TOOLS

Hot cut chisel

TECHNIQUES

Hot cutting

Tapering

Hot cutting is one of the magical techniques of the blacksmith's craft. There is something special about getting a solid piece of metal hot enough to cut through it like a piece of clay. After this project, celebrate your mastery of hot cutting by cooking up some hot steaks!

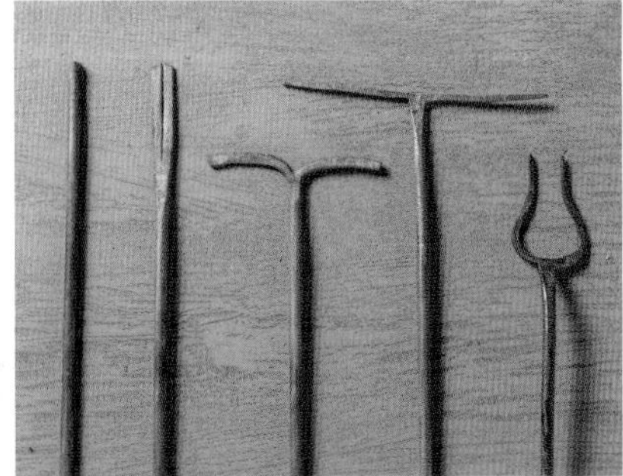

1. Prongs: Heat and flatten 2½" (65 mm) of round bar until it is ⅛" (3 mm) thick.

2. Hot cut 2" (50 mm) down the middle of the flattened end. ←

3. Taper each end of the split to a point. Fold one prong back so that the first can be worked, and then repeat for the other side. Make sure they are each tapered to the same length.

4. Clamp the piece into a vise just below the split. Use scrolling tongs or pliers to bend each prong out to about 90 degrees from the piece and then curve them back in line with each other. Offset the prongs slightly. You can also use the step or horn of the anvil for this process.

NEW TECHNIQUE: HOT CUTTING

Establish a cut line by centering the chisel at the end of the flattened section and striking down lightly. Move the chisel to the edge of the line created by the first blow and repeat until the full length of the line is created. This can be done at a lower heat. Working in this direction allows you to see the line you are creating. Reheat the piece and work back along the cut line with heavier blows to cut through the piece.

Place a piece of scrap sheet metal or copper plate under the piece to protect the anvil face and chisel (both hardened steels) from chipping. Remove the chisel from the cut after each blow to minimize heat transfer (which can affect the temper of the tool), and quench the chisel after every several hits for the same reason. Clean the base of the cut where the split occurs by spreading the ends and hitting the split against the end of the anvil.

HANDLE

1. Put a 2" (50 mm) taper on the end of the piece and then flatten 7" (18 cm) of the handle (including the taper) to ⅛" (3 mm) thick.

2. Optional step: Texture the flattened surface with the peen of the hammer.

3. Curl 1" (25 mm) of the tip to the underside of the handle.

4. Using bending forks or the anvil, create a gradual curve in the handle for an ergonomic feel. Wire brush and then finish the piece with a food-safe finish.

THE LEAF

BY JIM WHITSON, THE BLAZING BLACKSMITH
PEEBLES, SCOTLAND, UK

MATERIALS

½" (12 mm) round stock, run long

TOOLS

- Small flat stock tongs
- Scrolling tongs or pliers
- Flat chisel
- Fullering tool (optional)

TECHNIQUES

- Tapering
- Fullering
- Wrapping

Leaves are a fundamental design element in blacksmithing used in many projects throughout this book. Most smiths develop their own unique versions over time. Here are two different approaches to consider. After you get comfortable making leaves, try these same techniques with different starting stocks.

TIP: The neck can also be created by angling the stock against the corner of the anvil and striking angled blows against that spot. Positioning the piece this way causes the anvil and hammer to act as a fullering tool.

1. Put a short, ½" (12 mm) tapered point into the round stock.

2. Use a fullering tool such as a spring fuller or guillotine to create a shoulder, also known as a neck, 1¼" (30 mm) from the point. Do not make the neck too thin or the stem will break when working the piece.

→

3. Create a 3" (75 mm) taper in the base material.

OPTION A: CHISELED LEAF

1. Flatten the isolated section using heavy full-face blows. The piece spreads out much better when hot. Once the leaf shape is roughly formed, light, slightly angled blows at the edges can help even out the overall shape.

2. Use a chisel to make veins in the leaf. Start up the middle and work out from there. The leaf can be at a black heat for this step.

→

TIP: JUGGLING STEEL

Using chisels and punches requires the use of both hands. If working with long enough stock, free up your hands by holding the steel between your legs. Just make sure the metal is not too hot, or you may end up with a bad burn in a bad place.

←

3. Reheat and then dish the leaf using the step of the anvil, a curved cavity in a swage block, or a ground-out stump and then finish by curving the tip of the leaf over the edge of the anvil.

OPTION B: VEINED LEAF

1. Flatten the isolated section slightly using full-face blows.

2. Spread the leaf out using a fullering hammer. Do not hit in the center of the leaf, but rather work outward for each side, leaving a raised "vein" in the center. Repeat over several heats until the leaf is the desired width.

3. If you like the texture left by the fullering hammer, move on to the next step. To remove the texture, turn the leaf upside down and line it up the vein on the outside edge of the anvil. Use a combination of half-face and full-face blows to flatten out the ridges. Repeat for the other side of the leaf and then dish the leaf according to the process described for the chiseled leaf option.

KEYCHAIN

1. Cut off at 4" (102 mm) from the stem of the leaf.

2. Taper the stem to 3/16" (5 mm) in diameter and about 10" (250 mm) long.

3. Create a loop using the horn of the anvil. Flatten part of the loop to make it easier for adding jump rings.

4. Reheat. With the loop supported in a vise, wrap the remainder of the stem around the base of the loop using scrolling tongs or smooth pliers. Curling may take a few heats as the thin stem will lose heat fast. ↓

MAKING LEAVES WITH THE BLAZING BLACKSMITH

Sprouting bulbs, creeping ivy, and other natural forms inspire much of Jim Whitson's work where he has run the Blazing Blacksmith in his native Scotland for two decades.

While Jim has made thousands of leaves of all shapes and sizes in that time for hand-forged gates, railings, and sculptural pieces, the above styles are among his favorite for their simplicity and ability to make designs quickly come to life.

SKULL

BY DAVID GAGNE, ELM CITY VINTAGE
NEW HAVEN, CONNECTICUT, USA

MATERIALS

Hex nuts. The bigger the better. ¾" (20 mm) plain steel nuts with no zinc coating work nicely.

TOOLS

- Small tongs
- Ball punch
- Flat chisels

TECHNIQUES

- Flattening
- Spreading
- Using punches

What I love about this project is its creative use of an everyday material, hex nuts, and the potential for variety. Each skull can be unique, depending on your starting material. And with just a few punches and chisels, you can experiment with a bunch of different facial expressions.

CAUTION

Most nuts are galvanized in a zinc coating that creates a very toxic gas if heated, so make sure to either get plain steel nuts or remove the zinc coating before forging. This can be done with muriatic acid obtainable at your local hardware store.

1. Heat and flatten the nut to a uniform thickness. Spread the top half of the flattened nut wider than the bottom half using slightly angled blows or the fullering hammer until the piece has a pear shape.

2. Use a ball punch to make two eyes in the wider section.

3. Use flat chisels to make the mouth and nose. This can be done at a lower heat.
←

4. Optional step: Punch or drill a hole at the center of the skull or through one eye for a keyring.

5. Wire brush and finish.

HOOK

SUGGESTED MATERIALS

- ¼" to ⅜" (6 to 10 mm) round stock, run long
- ¼" to ⅜" (6 to 10 mm) square stock, run long

SUGGESTED TOOLS

- Twisting wrench
- Ball punch

TECHNIQUES

- Tapering
- Curling
- Twisting
- Punching

Even complicated designs rely on fundamental techniques, combined in creative and unique ways. Let's work through the process of layering techniques from past projects for this series of increasingly complex hooks.

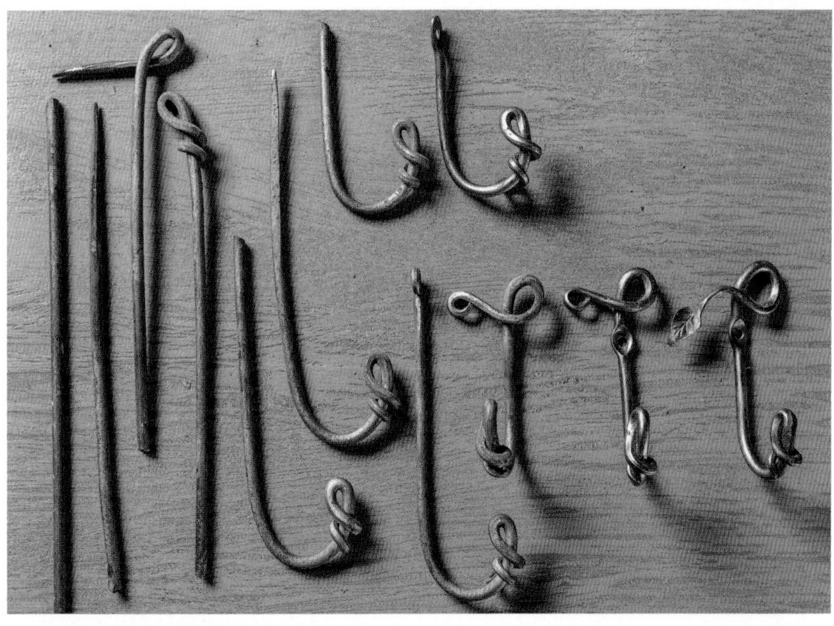

A. SIMPLE IVY HOOK

1. Forge a 3" (75 mm) round taper at the end of the round stock.

2. Curl 2½" to 3" (65 to 75 mm) back onto the piece using angled blows off the edge of the anvil and then flip the piece over to close the loop. Wrap the end using scrolling tongs or pliers (see Leaf project).

3. Create a larger curve, forming the "J" of the hook shape. Use light blows and move the piece after each blow to prevent kinks.

4. Measuring from the tip of the hook, cut off the remainder of the stock 2" (50 mm) from the tip.

5. Round the end and then offset the end using half-face blows with the tip of the hook facing downward.

6. Clean the end with a file or grinder, drill or punch a screw hole, wire brush, and finish the piece.

B. IVY-ER IVY HOOK

1. Follow steps 1 to 3 for Simple Ivy Hook. Cut off the remainder of the stock 5" (127 mm) from the tip and taper to 6½" (165 mm).

2. Forge a tight curl on the end (see Bookmark project) on a plane perpendicular to the plane of the J hook, and then create a larger loop on the same plane in the opposite direction. Make sure to leave 1" to 2" (25 to 50 mm) of space between the hook tip and the curls.

3. Use a ball punch (see Skull project) to create a depression for a screw hole, drill or punch a screw hole, wire brush, and finish the piece.

IVY LEAF HOOK VARIATION

Follow the steps for B, but forge a leaf on the end of the stock first.

C. SIMPLE HOOK

1. Forge a small square taper on one end.

2. Forge a tight curl on that end and then follow steps 3 to 6 from A.

D. TWISTY HOOK

1. After creating the larger curl of the hook, reheat and then clamp the back of the hook in the vise just below the tip.

2. Forge a tight curl on that end and then follow steps 3 to 6 from A.

NEW TECHNIQUE: TWISTING

Twisting wrenches are used to creates unique patterns in squared stocks. Making twisting wrenches is discussed in the Tools chapter, and varieties of twists are covered in the railroad-spike bottle opener project. A couple keys with twisting: having a two-handled tool will let you support the piece from both ends as you twist, preventing kinks. Metal will move where it is hottest, so make sure you have an even heat along the area being twisted. Finally, be sure to line up the untwisted ends after the twist is complete.

CALIPERS

BY NICHOLAS K. DOWNING, DOWNING ARTS, PORTLAND, MAINE, USA

MATERIALS	TOOLS	TECHNIQUES
½" (12 mm) round stock (rivet)	Riveting tool	Tapering
⅛" x ⅜" (3 mm x 10 mm) flat stock* cut to 5" (127 mm)	⅛" (3 mm) punch	Curving
		Punching
		Riveting

Caliper are a general term to describe a device used to measure the distance between two points. There are many variations of calipers, each designed for specialized uses. Two particularly handy for the blacksmith are "outside calipers" and "divider calipers," also known as a compass. Outside calipers allow the blacksmith to quickly reference measurements. A compass can be used the same way and can also be used for scribing lines and making circles.

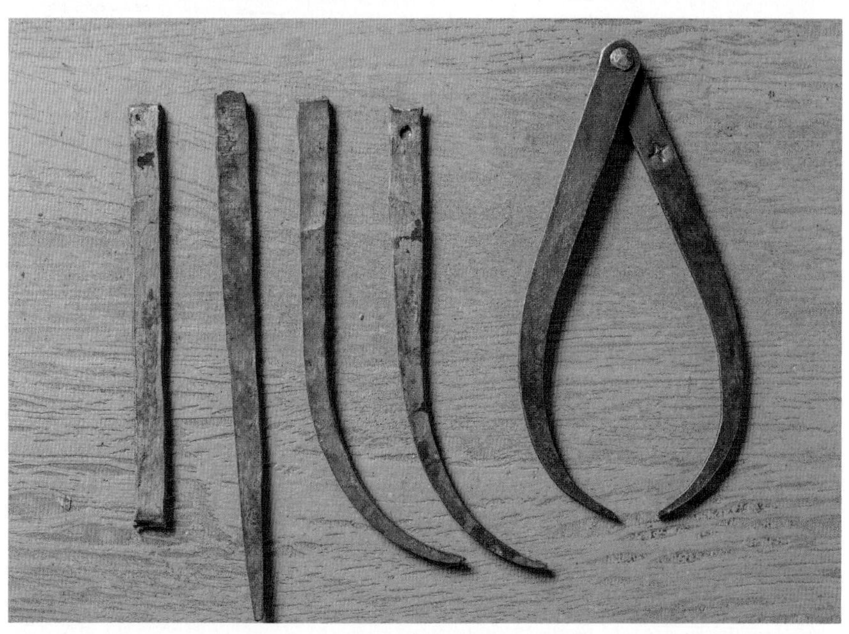

*As an alternative, use ¼" (6 mm) round stock and flatten to ⅜" (10 mm) wide.

1. Forge a 4" to 5" (102 to 127 mm) taper into each piece, maintaining the 1/8" (3 mm) width for a total length of 6" to 7" (150 to 178 mm).

2. Forge a curve into each point that increases its radius as it gets closer to the point. Skip this step to make a simple pair of dividers.

3. Grind or file the edges of the wide ends to a rounded shape and then punch or drill a 1/8" (3 mm) hole centered 1/4" (6 mm) from the ends of each piece. Clean up the holes and area around them with a file to ensure the calipers don't bind.

4. Rivet the piece together. If the rivet is too loose, cold peen slightly. If it is too tight, heat in the forge and manipulate the arms until they move more freely.

5. Move the arms to 90 degrees and file the profile of the ends round. Move them a few degrees closer and file the profile round. Continue the process until the points are touching.

6. Wire brush and finish.

NEW TECHNIQUE: HOLE PUNCHING

Hot punching holes is an alternative (and predecessor) to drilling. Rather than removing material like a drill does, a punch moves most of the hole's material out of the way, except for a small "slug" that is punched out to form the hole. This process makes the area around the hole stronger and can also create an aesthetically pleasing bulge around the hole. Get a good heat in the stock, line up the punch and hit down with the hammer. Repeat until you are almost through the stock. Quench the punch after every few hits to preserve the tool's heat treatment. Reheat, flip the piece over, and hammer the punch to complete the hole over the anvil's pritchel. You will see a slightly darkened bit on the underside where the thinned material has cooled faster. Use this as a reference to line up your punch.

NEW TECHNIQUE: RIVETING

Riveting lets you fasten pieces without welding and also make moveable joints such as for this project. I like to head my rivets before installing them using a rivet header (see Tools chapter). Head a rivet by upsetting 1½ to 2 times the diameter of the stock. First strike straight down to upset and then angled blows or the ball peen of the hammer can be used to finish the rivet. When completing the second end of the rivet, try to concentrate the heat on the end you are working to prevent the other end from continuing to upset. You can also use a spacer the width of rivet on the underside of the piece to minimize further upsetting.

BOTTLE OPENER

MATERIALS	TOOLS	TECHNIQUES
¼" (6 mm) round stock, run long	Small round tongs	Curves
$^{3}/_{16}$" (5 mm) round stock, run long (optional)		Twisting

Bottle openers are among the most popular small blacksmithing products today. Although I've made thousands over the years, I still have fun making them by approaching them like someone who paints might treat sketching. They are a way to warm up for the day's forging, practice hammer technique, and quickly explore new designs for handles, subtle variations of curves, and other techniques that can later be incorporated into larger pieces.

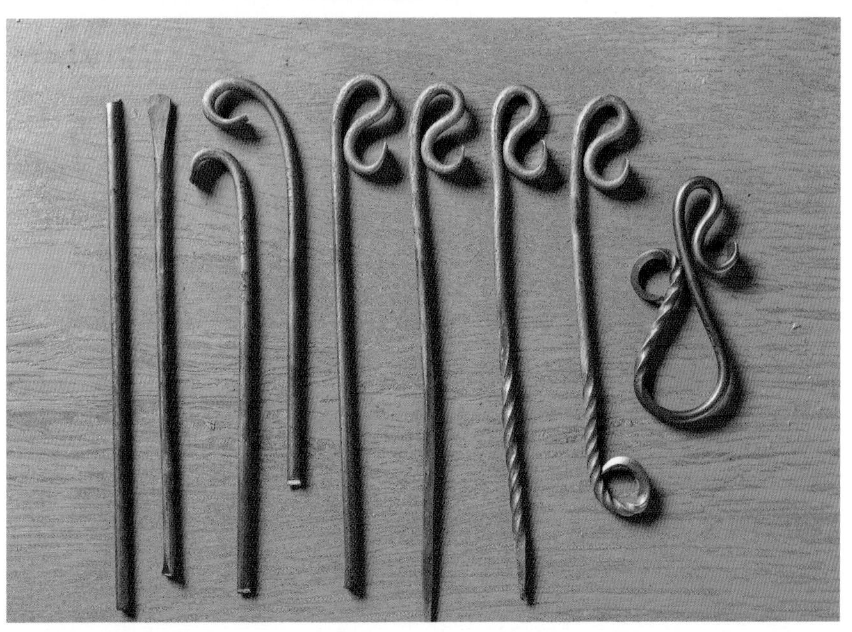

1. Use angled blows to make a screwdriver tip at the end of the stock. Curl the tip into a "C" between ½" to ¾" (12 to 20 mm) in diameter using angled blows over the bevel of the anvil and then backward facing blows to finish the "C."

2. Use a small bending fork secured in a vise to create the second and third bends that complete the opener head.

NEW TECHNIQUE: USING SCROLLING FORKS

Steel naturally bends where the combination of heat and force is greatest. Scrolling forks can be used singly or in pairs to precisely control where force is applied in bends and, therefore ,where bends occur. Tip: To keep the bends tight, pull the stock away from the fork as you bend.

3. Optional step: Flatten the outer curve to create a space for personalization.

4. Cut off the end 8" (200 mm) as measured from the beginning of the "head" of the opener and forge a 6" (150 mm) square taper in the end.

5. Heat and quench the first 1" (25 mm). To create a twist in the squared off section, secure the point in a vise and use a twisting wrench. Because the material gets thinner toward the tip, it will twist more, creating an irregular pattern.

6. Forge a 1" (25 mm) loop facing the same direction as the opener. Quench the loop, then forge a larger loop to close the handle. Work the loop around the horn to make a teardrop-shaped curve.

VARIATIONS

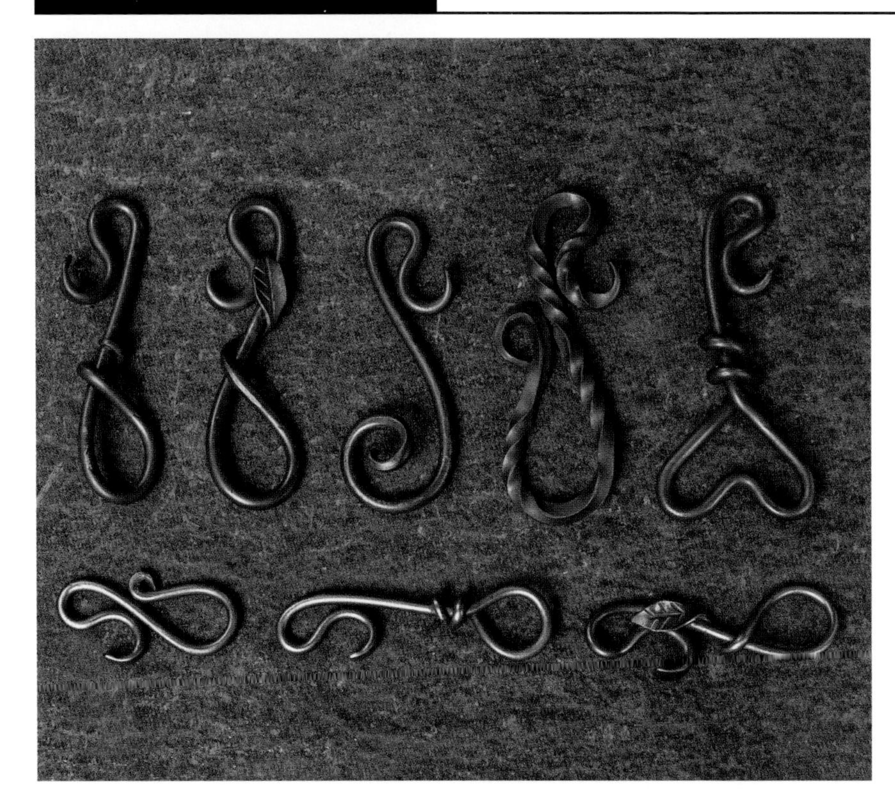

There are countless possible variations with the same style opener top. See if you can make some of these based on the cut-off lengths from step 4 above.

A. Wrap: Cut at 7" (180 mm)

B. Leaf: Cut at 8" (200 mm)

C. Pig Tail: Cut at 7" (180 mm)

D. Reverse Twists: Use ¼" (6 mm) square stock, cut at 8" (200 mm)

E. Heart: Cut at 12" (300 mm)

Key Chains use with 3/16" (5 mm) stock

F. Mini key chain: Cut at 3½" (90 mm)

G. Loop key chain: Cut at 6" (150 mm)

H. Leaf key chain: Cut at 6" (150 mm)

RUSSIAN ROSE

SUGGESTED MATERIALS

5/16" (8 mm) round stock cut at 14" (356 mm)

SUGGESTED TOOLS

Scrolling tongs

TECHNIQUES

Tapering

Curling

Steel, more than any other material, epitomizes man's mastery of the natural world. However, replicating delicate natural forms in this material requires a different kind of mastery. Of all the starting projects, this one requires the most "feel" to make the piece come alive. See how well you can emulate nature with this rose project.

ROSE

1. Flatten the first 5" (127 mm) of the bar slightly and use the hardy to cut ¾-way through the narrower width at approximately ½" (12 mm) intervals. The initial marks can be made cold.

2. Flatten the 5" (127 mm) section with a taper so that the edge cut into (the top of the petals) is thinner. Leave the last petal before the stem slightly thicker.

←

3. Bend the petals to a 90-degree angle from the stem, and then use the peen of the hammer on the step of the anvil to begin wrapping the petals around the stem. Finish curling the petals by lightly hammering and turning the piece onto itself. This may take several heats.

→

4. Get a good heat into the petals, clamp the stem into a vise, and use scrolling tongs or pliers to pull down and shape the petals. Don't work the petals too cold or they can break.

STEM

1. Taper the end to a point and curl into a tight loop. Make an additional curl until there is approximately 2" (50 mm) of straight stem remaining.

2. Clamp the curved section of the stem in the vise and use tongs to align the rose over the center of the curve. Adjust as necessary to make sure the piece stands on its own.

3. Wire brush and finish. A brass brush can be used to give the petals a nice accent.

PART II:

PROJECTS

PROJECTS

CHAPTER 4: TOOLS

This chapter focuses on the everyday tools that will greatly assist you in making the rest of the projects within this book. Being able to make tools is one of the primary skills that distinguishes blacksmiths from other craftspeople. While others make products, we are able to make the tools that enabled those products to be made in the first place. Does that also mean we are better than other tradesfolk? Perhaps. More humble? Definitely.

A few of the tools in this chapter involve heat treating, a process used to harness qualities of strength and durability in special steels called "tool steels," also known generally as "high-carbon steels." The ability to heat treat and temper steel centuries before modern metallurgy explained the process was another attribute that elevated the blacksmith's renown within the community. The smith's power to quench a piece of razor-sharp steel to make it hold an edge seemed a form of mysticism or magic. Even today, with full knowledge of the process, heat treating a tool and then seeing its properties completely transform still seems magical.

BENDING FORKS

SUGGESTED MATERIALS

- $\frac{5}{16}$" (8 mm) round stock cut at 14" (356 mm)
- Small forks: $\frac{3}{8}$" x 1" (10 mm x 25 mm) run long
- Medium forks: ½" x 1¼" (12 mm x 30 mm) run long

SUGGESTED TOOLS

- Handheld fuller

TECHNIQUES

- Offsetting
- Drawing out
- Fullering

Bending or scrolling forks are an indispensable tool for the blacksmith. If you have access to a welder, they can be quickly made by welding two round stock "tines" onto a square stock handle (square stock handles will seat better when clamped in a vise). You can also make a simple fork for your vise by creating a "U" with a piece of round stock. However, if you want to make them the old-fashioned way, try this project.

NOTE: Project steps are written for the medium forks with any modifications for the smaller forks appearing in brackets.

1. Forge a shoulder 1¼" (30 mm) [1" (25 mm)] from the end of the piece using half-face blows on the near side of the anvil. Draw the shoulder out to ½" (12 mm) [3/8" (10 mm)] square.

2. Forge another shoulder 1" (25 mm) from first shoulder using half-face blows on the far side of the anvil. Make the section the same dimensions as the first shoulder.

3. Bend the piece into a "C" shape with the isolated section facing out from the center of the curve and forge this section into the second "tine," matching the dimensions of the rest of the fork. A fullering chisel or hammer is useful to direct these blows.
←

4. Lightly hammer out the curve. Round each tine slightly with the hammer.

5. Curve the end into the final fork shape. A ¾" to 1" (20 to 25 mm) gap is a good general size for the medium forks and a 3/8" to ½" gap (10 to 12 mm) works well for the smaller forks.

6. Cut or grind the tines so they are the same length and round the tines with a file or grinder.

7. Draw out the handle to 16" (400 mm) [13" (330 mm)] including a 3" (75 mm) taper. Leave the first 6" to 7" (150 to 178 mm) squared off and round the remainder.

8. Forge a curl on the end of the handle in the opposite side as the forks.

CHAIN HOLD DOWN

SUGGESTED MATERIALS

- Used motorcycle chain or similar (such as #50 roller chain)
- ¼" x 1" (6 mm x 25 mm) flat stock cut to 24" (610 mm)
- ¼" (6 mm) round bar, run long

SUGGESTED TOOLS

- Medium bending forks
- Small round tongs

TECHNIQUES

- Bending with forks
- Punching holes (optional)
- Riveting (optional)

The chain hold down is great all around "third hand" helper; it is very quick to set and release and it will hold nearly any size piece at any point on the anvil. Old chain can sometimes be obtained free at motorcycle shops. When attaching the stirrup, hook the chain hook into the last motorcycle chain link and attach the end of the chain to the anvil base. The chain should be attached so that when placed across the bare face of the anvil, the stirrup is approximately 1½" (40 mm) off the ground.

STIRRUP

1. Drill or punch out each end of the flat stock ½" (12 mm) from the end, making the holes slightly larger than ¼" (6 mm).

2. Mark the stock at the points A, B, C, and D as indicated in the diagram and bend each mark to create the final triangle. Start with marks A and D, bending them in the same direction and then bend marks B and C in the opposite direction as A and D. Tweak as necessary to make sure the holes line up with a ½" (12 mm) gap between the holes.

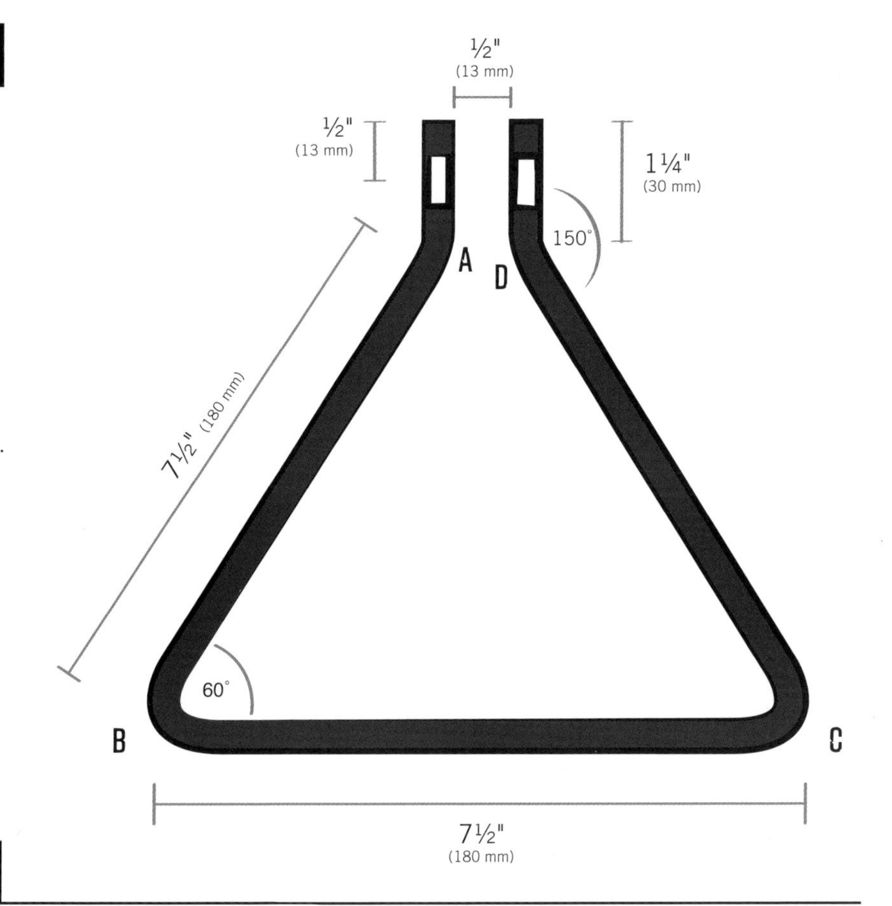

CHAIN HOOK

1. Forge a loop with a 1" (25 mm) diameter out of the ¼" (6 mm) round stock.

2. Cut off 1½" (40 mm) from the beginning of the loop and forge a 1" (25 mm) taper.

3. Curve the taper into a hook shape with a 1" (25 mm) diameter, leaving approximately ⅜" (10 mm) of the hook open.

ASSEMBLY

OPTION 1 (SIMPLE):

Use a ¼" (6 mm) bolt to attach the chain hook to the stirrup.

OPTION 2 (MORE COMPLEX):

Use a ¼" (6 mm) rivet. Be careful when making the second rivet head not to pinch the stirrup and the chain hook.

TWISTING WRENCH

SUGGESTED MATERIALS

- ½" (12 mm) square stock cut to 34" (864 mm)
- ⅜" (10 mm) square stock cut to 20" (51 cm)
- $^{3}/_{16}$" (5 mm), ¼" (6 mm), ⅜" (10 mm), and ½" (12 mm) spacers cut to 12" (300 mm)

SUGGESTED TOOLS

- Small and medium bending forks

TECHNIQUES

- Bending
- Tapering
- Curling

If you have already made the hold down and bending forks projects, you can put both to use to make these twisting wrenches.

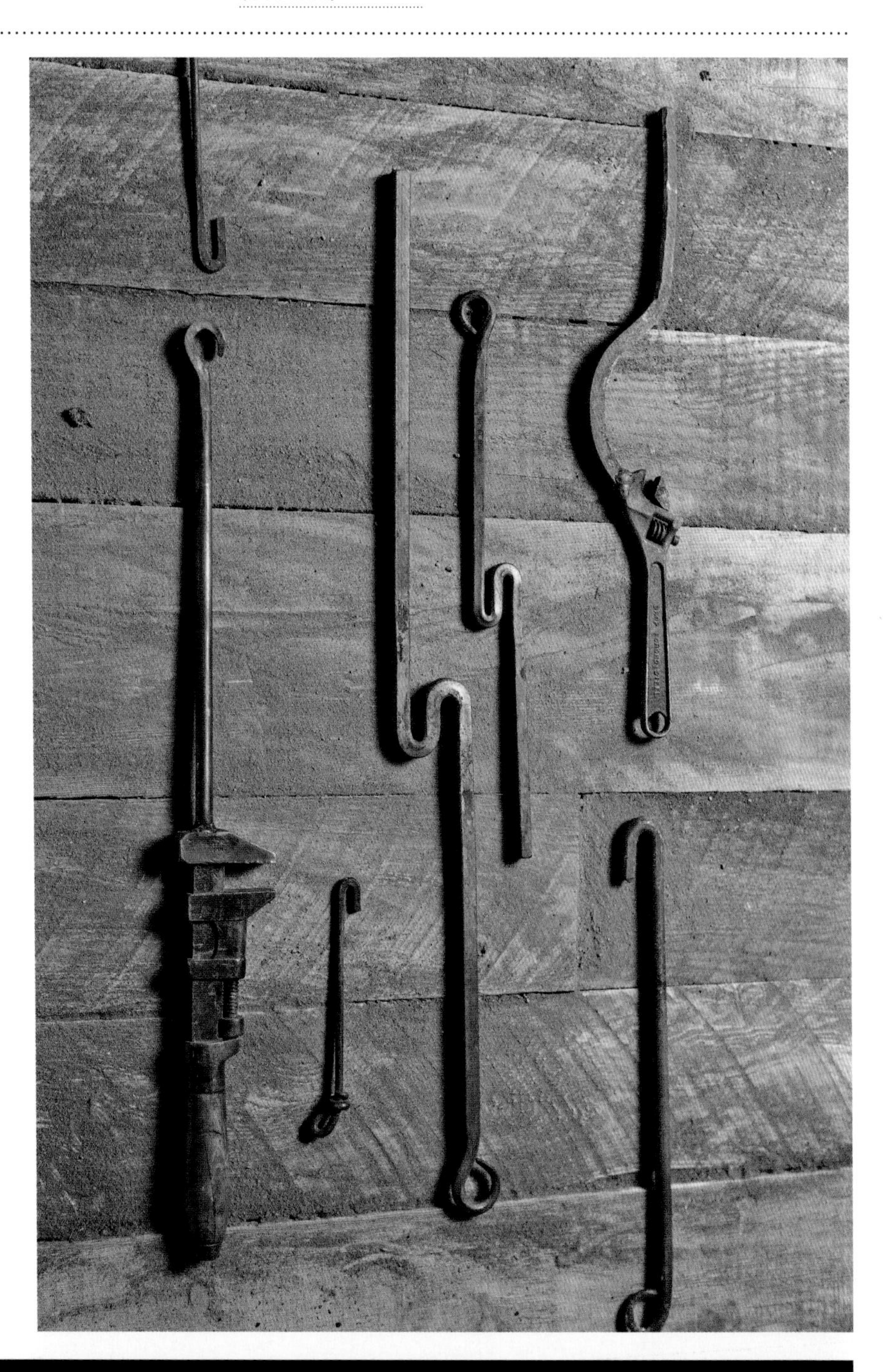

TIP: An adjustable twisting wrench can be made by welding a second handle onto a monkey wrench.

NOTE: Project steps are written for the ½" (12 mm) stock wrenches with modifications for the smaller wrenches appearing in brackets.

1. Bevel the edges of the center 12" (300 mm) of the stock slightly.

2. Mark the center of the square stock and heat that section. Use bending forks to get a tight 180-degree curve starting 1" (25 mm) from the center mark. Insert the ½" (12 mm) [¼" (6 mm)] spacer along the length and flatten the wrench to fit that spacer.

3. Reheat the center, quench the first curve, and then create a second tight curve mirroring the first to create an "S" shape.

4. Reinsert the first spacer into the original curve and the ⅜" (10 mm) [3/16" (5 mm)] spacer into the new curve and flatten the curves to fit each spacer. Support one spacer with your hand and the other with the hold down.

←

5. Tweak as necessary to achieve a secure but not too tight fit and make sure the "arms" are parallel.

6. Forge a 5" (127 mm) taper on one end and create a small loop for hanging the wrench.

HEAT TREATING

MATERIALS	TOOLS	TECHNIQUES
High-carbon steel	Torch (oxy-acetylene)	Annealing
		Hardening
		Tempering

Tools such as chisels and punches need to be harder and stronger than the steels they are used to work. Therefore, these tools are made of high-carbon steels known collectively as "tool steels." The higher carbon content of tool steels allows the steel to be heat treated, a process first used by blacksmiths in Turkey more than 3,000 years ago.

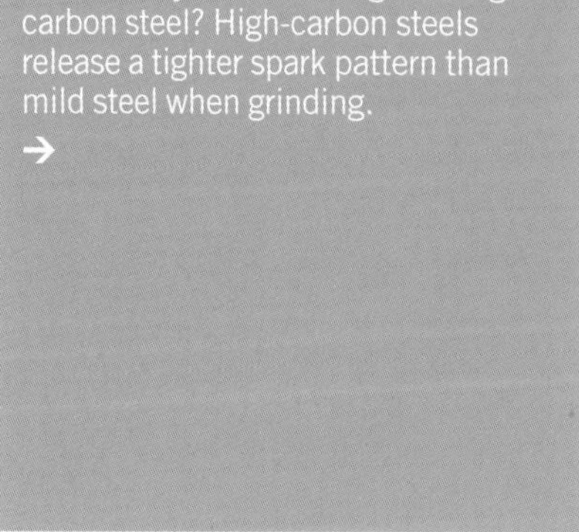

Not sure if you are dealing with high-carbon steel? High-carbon steels release a tighter spark pattern than mild steel when grinding.

→

Heat treating steel consists of three steps: *normalizing* or *annealing*, a process that softens the steel and removes internal stress; *hardening*, a process that maximizes the hardness of the steel but makes the piece brittle; and *tempering*, a process that toughens the piece and undoes brittleness.

1. FORGING

Each tool steel has different temperature ranges for forgeability, but generally speaking you need to work the piece more carefully than mild steel to prevent stress and cracks from building up in the piece. Try to not work the piece hotter than a bright orange or colder than a dull red.

2. ANNEALING

Again, specific temperatures will vary by steel, but annealing consists of heating the piece to around cherry red and then letting it cool as slowly as possible.

3. HARDENING

Heat the tool to medium red or the point that it becomes nonmetallic and then quench in the proper medium (water or oil). Some smiths keep a magnet handy to determine the proper hardening temperature.

4. TEMPERING

Polish the piece to the bare metal, heat slowly until the working end picks up the appropriate color for intended use (see chart at right) and quench. Test the temper by running a file over the working end. If the file glides over the surface, the piece has been hardened and tempered.

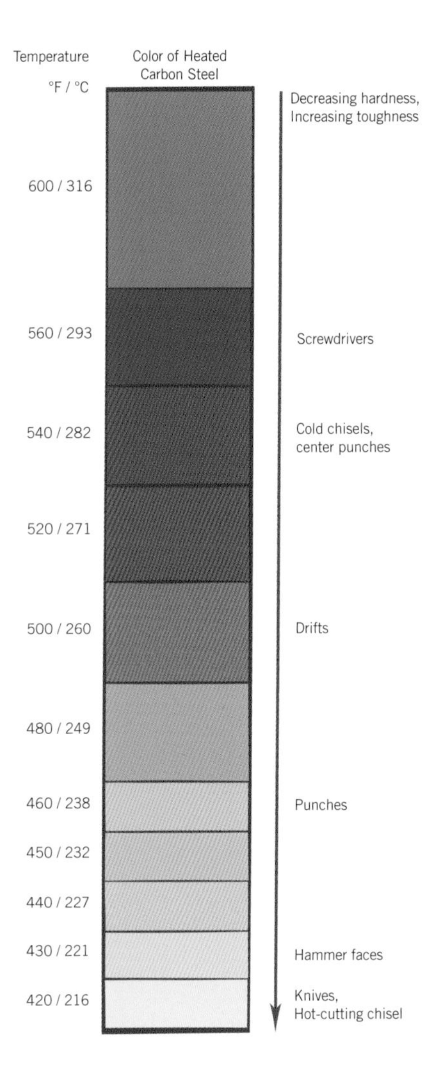

STARTING STEEL: BUYING NEW VERSUS SALVAGED

Salvaging tool steel has the benefit of being very cost effective. The downside is that it requires a greater investment of time to determine the type and quality of steel with which you are working. If using salvaged pieces, you will have to experiment to determine the best quenching medium. If buying new, there are many different types of tool steels to choose from, depending on your purposes. They are generally classified by the medium in which they are hardened: water, oil, or air. W1 (water quenched) and 4140 (usually oil quenched) are common choices and good for all-around use. Experienced smiths will tell you to learn how to use a particular steel well and stick with it.

PUNCHES, CHISELS, AND DRIFTS

SUGGESTED MATERIALS

½" (12 mm) W1 drill rod or other high-carbon steel

¾" (20 mm) W1 drill rod or other high-carbon steel

SUGGESTED TOOLS

Rasp or grinder

Specific projects require specific tools, even throughout this book. The process for making most tools is similar, regardless of the final tooling ends. These steps cover basic tooling concepts and can be applied to larger or smaller more specialized tools.

TECHNIQUES

- Heat treating
- Tapering
- Grinding/filing

HANDLES

Forge a slight taper in the struck end of the tool and round the top with a rasp or grinder. The taper keeps the piece from mushrooming, and the rounded top helps the tool still hit downward when mishit slightly.

CHISELS

Forge a flat taper into the working end of the piece and then finish the edge with a rasp or grinder. For hot-cutting chisels, make the final edge 30 degrees and leave the working end slightly rounded. For cold-cutting chisels, make the final edge 60 degrees. For fullering chisels, round the with a rasp or grinder.

PUNCHES

Round punch: Forge a rounded taper into the working end to the desired diameter, then grind or file the end flat.

Slot punch: Forge a flat taper to the desired side, then grind or file the end flat.

Center punch: Follow the process for a round punch, then forge or grind a 60- to 90-degree tapered point on the end.

DRIFTS

Once a hole is punched, drifts are used to expand the hole into the desired size. Drifts can be made of mild steel. Start with stock of the desired hole size and type. Taper the hammered end slightly and make a longer taper on the working end.

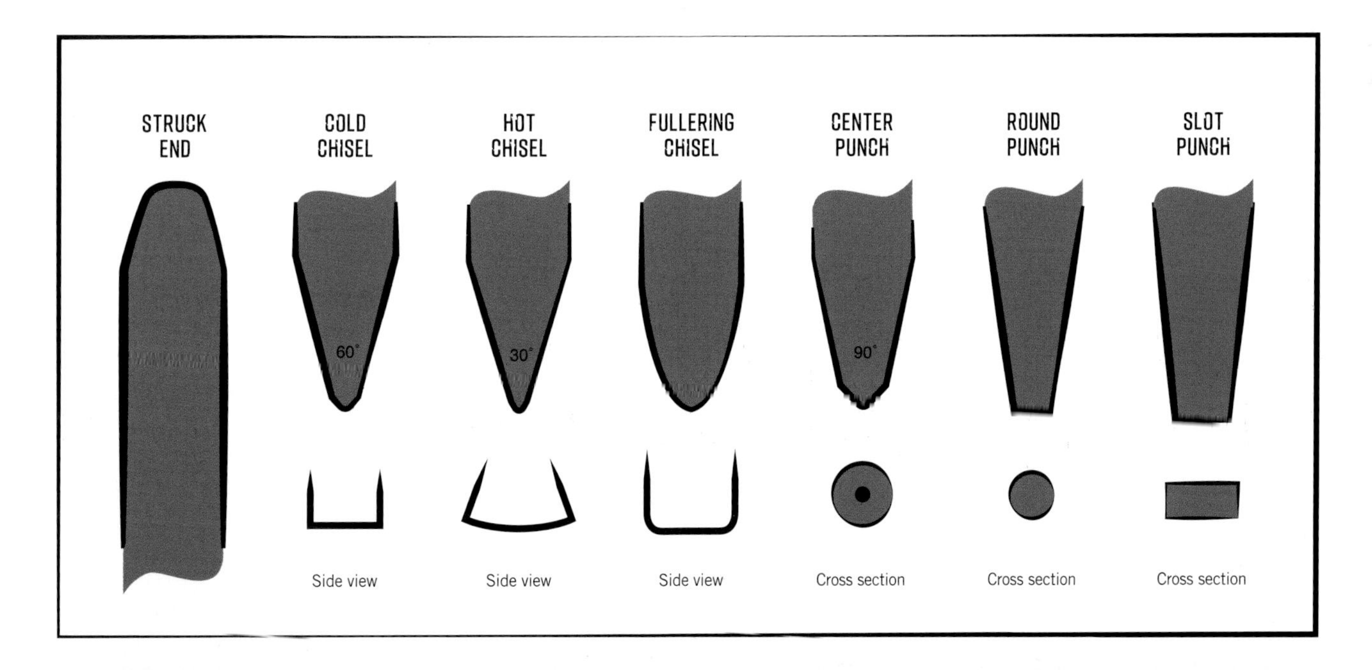

RIVET HEADER

SUGGESTED MATERIALS

¾" (20 mm) square stock run long

SUGGESTED TOOLS

Fuller

Clamp

TECHNIQUES

Offsetting

Curving

Grinding

I like making rivets in a rivet header so I can run them long and cut them to the required length when it's time for assembly. Holes in the header are made slightly smaller than the rivet stock. The rivet stock is placed into the header, and a vise is used to clamp the rivet together while the head is made. I like to use a torch when making rivets, but you can also use the forge if you work quickly.

1. Mark the square stock at 6", 8", and 14" (150 mm, 200 mm, and 350 mm).

2. Heat the center up and use a fuller to isolate the section between 6" (150 mm) and 8" (200 mm). In a pinch, the edge of the anvil or a scrap piece of round stock can be used as a fullering tool.

↗

3. Draw out the isolated section to 8" (200 mm), keeping the width the same as the parent stock.

4. Get an even heat in the middle section and create a loop to bring the ends together. This can be done with scrolling tongs or on the anvil. While it is okay if the ends are slightly offset along their lengths, make sure they are exactly lined up along their height. Cut the excess stock.

5. Optional step: Grind or chisel out a lip along each outside bottom length approximately 3/8" (10 mm) into the piece and 1/2" (12 mm) up from the bottom. This will prevent the header from "sinking" under hits in the vise.

6. Clamp the piece tightly with a C-clamp or similar and drill out holes slightly smaller than desired rivet sizes along the center of the piece.

WOLF JAW TONGS

BY FININ LIAM CHRISTIE,
TRADITIONAL BLACKSMITH, GOREY,
COUNTY WEXFORD, IRELAND

SUGGESTED MATERIALS

- ¾" (20 mm) round stock cut to 8" (200 mm)—Stock can be run long and cut to length last to avoid needing tongs.
- ⅜" (10 mm) round stock for rivet
- ¾" (20 mm) fuller or round stock run long
- ⅝" (16 mm) fuller or round stock run long

SUGGESTED TOOLS

- ⅜" (10 mm) punch

TECHNIQUES

- Punching
- Riveting
- Drawing down
- Offsetting
- Fullering

Wolf jaws are the Swiss army knife of tongs. They are a great generalist tong for holding multiple-sized stocks in multiple directions. These steps make tongs that are meant for holding ½" to 1" (12 to 25 mm) round stock, but they can be modified to hold larger or smaller stocks.

1. Mark the stock at 1" (25 mm) and 2" (50 mm). Follow these steps for both sides of the tongs.

2. Use half-face blows on the near side of the anvil to isolate the first 1" (25 mm) of stock. These will become the "jaws."

3. Position the jaws sideways off the back edge of the anvil angled away from your hammer hand about 45 degrees. Use half-face blows to define the first half of the "boss" (center portion).
→

4. Define the back of the boss with half-face blows off the back of the anvil on the opposite side of the jaws at the original 2" (50 mm) mark.

5. With the underside of the jaws facing up on the anvil, use a piece ¾" (20 mm) round bar perpendicular to the jaws as a fuller to begin to define the first groove.

6. Define the next groove by starting ¼" (6 mm) from the end of the first groove. It is important to leave enough space between each so that they match when fully formed. Continue to work the grooves until they are both fully formed.

7. Use a piece of ⅝" (16 mm) round bar perpendicular to the formed grooves to create a channel down the middle of the jaws.
↑

8. Draw out the "reins" to the desired length and thickness.

9. Punch a ⅜" (10 mm) hole in the center of each boss for the rivet. Rivet the tongs together using ⅜" (10 mm) rivets.

10. Adjust the jaws and reins as necessary for proper fit and action. If the rivet makes movement too tight, heat the boss up and open and close the reins until the tongs move freely.

IN THE SHOP WITH FININ LIAM CHRISTIE

Finín has been blacksmithing for more than 40 years in his native Ireland. In that time, he's made a lot of tongs. If you find yourself making a few yourself, here is a special jig Finín uses for steps 2 to 6 that you can make with access to a welder.

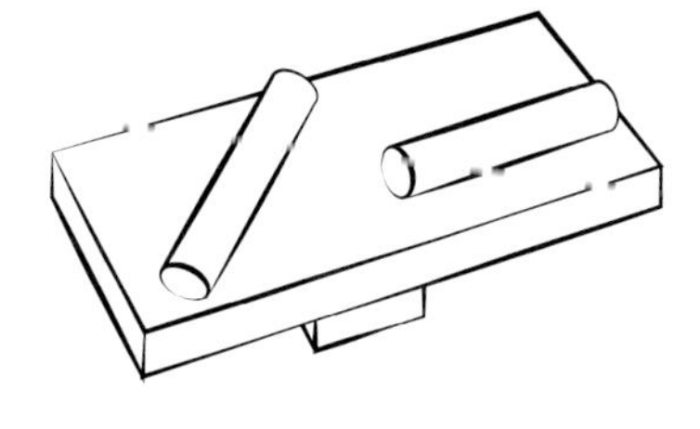

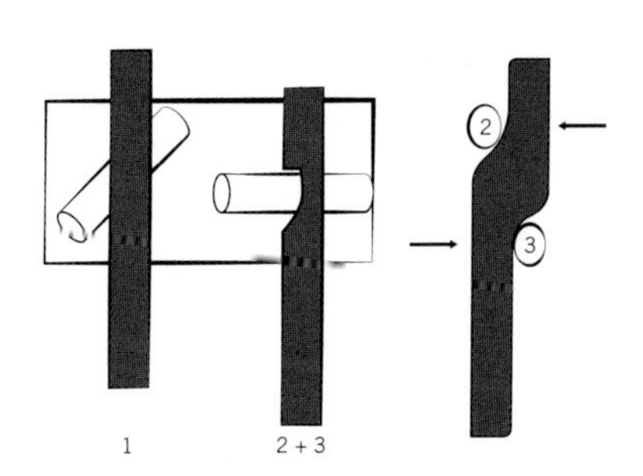

PROJECTS

CHAPTER 5:

AROUND THE HOME

I learned about the effect that handmade pieces can have on our everyday lives working in my dad's basement as a kid. On rainy weekends, he and I would piece together scraps of wood to make toy ships, tanks, and old-school cannons. I still have one of our little boats in my office today. It is almost as if that boat is a talisman with the power to transport me back in time. One look at it and I can go right back to those weekends with dad.

In my time metalworking, I've found that experience is not unique. When we make things with our own hands, we do more than transform raw materials, we also add a bit of ourselves to the piece. I think that is one of the reasons it feels so good around the home to surround ourselves with handcrafted objects like the projects in this chapter. Each piece, no matter how small, tells its own story, a story that grows along with us every time we use it.

LETTER OPENER

SUGGESTED MATERIALS

¼" (6 mm) round stock, run long

SUGGESTED TOOLS

Small round tongs

TECHNIQUES

Tapering

Flattening

Handles

Opening actual mail instead of email has a nostalgic feel that pairs nicely with a handmade letter opener. You can personalize the blade easily using a stamp set—I've had good luck with 3/16" (5 mm) and ¼" (6 mm) stamps. You can also create many handle variations for added customization. These handle designs, in turn, can be applied to larger projects such as fire pokers.

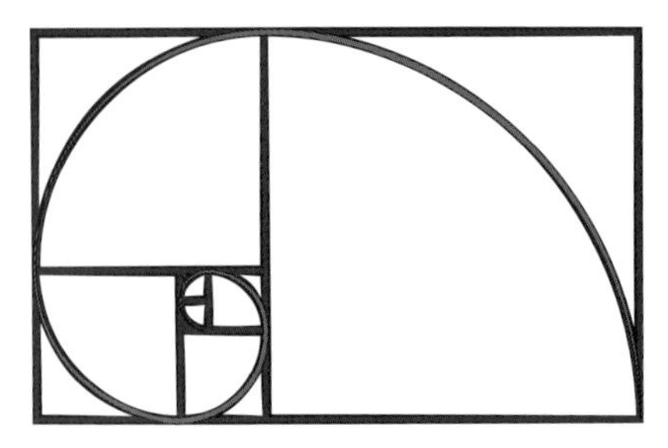

NEW TECHNIQUE: MAKING SCROLLS

Scrolls are one of the most common designs in traditional blacksmithing. They can be made from many different starting materials, with various ending tips, and spiraling patterns. A traditional pattern loosely follows the golden ratio, in which every succeeding point on the scroll is farther from the inner spiral preceding it.

Start by working the tip over the horn or beveled edge of the anvil. When a "C" shape has been formed, flip the piece over and use backward blows to continue curving the spiral, adjusting the curve with the location of the blows and angle of the stock. Work the piece back and forth in this manner until the desired spiral is achieved. Having a nice, even heat will help create smooth curves. Scrolling forks can be used to adjust kinks and assist with larger scrolls.

HANDLE

Forge a 1" (25 mm) taper on one end and scroll the end to about 1½" (40 mm) in diameter. Once the scroll is complete, forge a tight bend on the shape in the opposite direction of the scroll to center the scroll in relation to what will become the blade. See new technique.

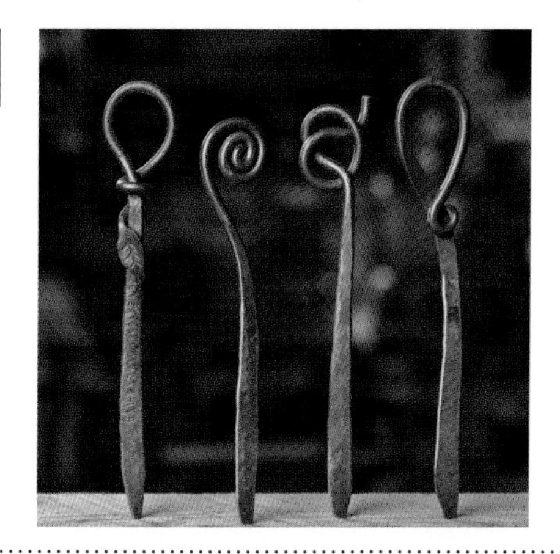

HANDLE VARIATIONS

A. IVY LEAF

Forge a leaf on the tip (see Leaf project). Make about an 80-degree bend starting 8" (200 mm) from the leaf and then curl back into a loop with about 3" (75 mm) of overlap. Before wrapping the leaf, follow steps for making a blade.

B. SPIRAL

Start by rounding the end of the stock and hammer tight curls over the edge of the anvil. After several spirals, make a sharp bend in the opposite direction, then make the blade.

C. KNOT

Make a large loop, about 4" (102 mm) in diameter, with the tip overlapping the loop by several inches. Use tongs to fold the tip into the loop and then use a combination of hammer blows and tongs to tighten the "knot." Cut off the excess, then make the blade.

D. DOUBLE CURLS

Taper the end into a small point. Create a tight curve and then a larger loop in the opposite direction, but don't close the loop fully. Follow the steps for making a blade. Finally, use scrolling tongs or pliers to center the first curl onto the blade.

BLADE

1. Cut off at 6½" (165 mm) from where the handle meets the shaft and put a 1" (25 mm) taper on the end.

2. Flatten the blade to approximately ½" (12 mm) wide. Blend the transition between the handle and the point so that the opener is at its widest right before the point starts.

3. Bevel the blade with angled blows along each edge of the blade and create a sharper edge with a file or grinder. Wire brush and finish.

CLOVER LEAF AND HOOK

BY MATT JENKINS, CLOVERDALE FORGE
SAINT ANDREWS, MANITOBA, CANADA

SUGGESTED MATERIALS

- ½" (12 mm) square stock, run long

SUGGESTED TOOLS

- Fullering tool
- Handheld fuller

TECHNIQUES

- Fullering
- Offsetting
- Spreading
- Punching
- Curling

Hooks are a such a common element of blacksmithing that their creative potential can be easily overlooked. However, they are a great starting point for exploring new designs and techniques, the only constraint being your own imagination. As an example, this clover leaf hook was one of a series of 366 unique hooks made by Matt Jenkins over the course of a calendar year. Try this one out, and then see if you can innovate a few of your own design.

CLOVER LEAF

1. Line up the stock approximately 3/8" (10 mm) onto the anvil and use half-face blows to forge a rounded tab at the end of the stock.

2. Fuller the two sides of the piece approximately ¾" (20 mm) from beginning of the tab and taper slightly to "neck" the piece. Then use half-face blows to offset the top of the isolated section and continue the taper further.

3. Use a small fuller to divide the isolated mass of steel in the middle along its length, then flatten all three masses into the rough clover shape. ↑

4. Use a ball punch or similar rounded tool to add some dimension to the clover leaves, then use a file or grinder to complete the final clover shape.

HOOK

1. Cut the stock 3" to 5" (75 to 127 mm) from the start of the clover and forge the ½" (12 mm) stock down to approximately 3/8" (10 mm) square.

2. Forge a square taper into the end, curl the tip in the opposite direction as the top of the clover, quench the tip, and then make a larger curl to complete the hook shape.

3. Punch or drill out a hole below the clover. Wire brush and finish.

SUGGESTED MATERIALS	SUGGESTED TOOLS	TECHNIQUES
Old horseshoes	Flat stock tongs	Tapering
		Curling

DOUBLE-HOOK HORSESHOE

BY DERECK GLASER, DG FORGE
WINTHROP, MAINE, USA

There's a whole branch of blacksmithing that deals with the art of working with salvaged materials. Besides being inexpensive, salvaged pieces offer a chance to branch out from the uniformity of metal stocks. This horseshoe project is a nod to that mindset, as well as to the blacksmith who originally made that shoe.

IN THE SHOP WITH DG FORGE

As metal objects age, rust, and wear down, their surfaces develop beautifully nuanced patterns betraying the piece's unique history. Dereck Glaser loves working with salvaged steel to highlight such features, as with this vessel made from pitted tubing.

1. If the shoe has clips, forge those flat to the backside of the shoe.

2. Heat the center of the shoe (the "toe"), clamp one side (a "branch") in the vise and twist the other branch 180 degrees from its original position.
→

3. Forge an approximately 6" (150 mm) taper into one branch. Record the exact length with a compass, using one of the shoe's nail holes as a reference for repeating on the other branch.

4. Forge a small scrolled end backward, then bring the full hook forward.

5. Repeat for the other branch. The hooks will seem to be in opposite directions at this point.

6. Get an even heat in the toe and twist the piece back into the original position.

7. Adjust if needed to make sure the hooks are symmetrical and hang evenly.

8. The hooks can be hung using the existing nail holes, but if larger screw holes are desired, drill or punch them at this point. Wire brush and apply finish.

BOWL

BY CHRIS DANBY, COACH HOUSE FORGE
CHESHIRE, ENGLAND, UK

SUGGESTED MATERIALS

- 1/8" (3 mm) mild steel plate at least 10" x 10" (250 x 250 mm)
- 8 gauge copper wire

SUGGESTED TOOLS

- Plasma cutter, torch, or angle grinder
- Hot chisel
- Swage block or former
- Small ball-peen hammer

TECHNIQUES

- Hot cutting
- Forming
- Rivets

Properly executing seemingly simple design choices can start to elevate and differentiate a piece. With this bowl, overlapping steel ends and copper accents transform a simple vessel into an heirloom-quality home good. If you don't have access to a swage block when forming the bowl, a former can be made by grinding out a depression in a tree stump or wood block.

IN THE SHOP WITH COACH HOUSE FORGE

Since starting Coach House Forge in 2014, Chris Danby has developed a wide range of bowls and vessels with wonderfully unique designs. Using different steel forms, varying textures, and stamping personalized messages are just some of the ways he elevates his bowls. One piece of advice he has for making bowls is not to rush the dishing process, but, rather, dish gradually over several heats.

1. Mark and cut out a 10" (250 mm) disc from the stock material and clean out any burrs with a file.

2. Mark the center of the disc and make a straight line from the center to the edge. Mark the line with a center punch along 1" (25 mm) intervals.

3. Heat the disk and cut along the punch marks. Gently open the "cut" on the anvil and dress the edges with a file. Heat and reclose the circle.

4. Heat the disc and begin to form the bowl in a swage, using an 8" (200 mm) diameter swage if you have access to one or grind one out of a stump. This will take several heats. Start forming around the outer edge of the disc hammering in ever decreasing circles. Take care to ensure that the cut ends form an overlap and that the bowl shape is maintained.
↑

5. Form a "flat" at the base so the bowl sits upright.

6. Wire brush the piece.

7. Center punch and drill out three holes for rivets. Cut the copper rivets to size and set them in, ensuring a tight fit.

8. Lightly wire brush and finish the piece.

BOOKENDS

SUGGESTED MATERIALS

$\frac{3}{16}$" x 3" (5 mm x 75 mm) flat stock

2 cut to 30" (750 mm)

SUGGESTED TOOLS

Scrolling tongs

Bending forks

TECHNIQUES

Upsetting

Folding

Curling

Metalwork at its best follows clean lines and smooth curves. These bookends are a simple design but require careful execution to achieve that graceful final shape. See how symmetrical you can make a set, or make them "intentionally" asymmetrical.

1. Hot cut or grind a 45-degree cut into one end of the stock.

2. Starting from the flat end, mark the flat stock at 7", 13", 18", and 23" (178 mm, 330 mm, 457 mm, and 584 mm).

3. Upset the flat end slightly.

4. Begin creating the folds working from the upset end. This can be done using the edge of the anvil or large bending forks. Each fold will be in the opposite direction of the previous fold. Use your marks as a reference for where each fold should be centered and quench the already folded sections of the piece as needed to prevent distortions.

→

5. For the last 7" (178 mm) containing the tapered end, work the piece over the edge of the anvil at a 45-degree angle to curl the tapered end onto itself slightly. Use scrolling tongs to finish the last fold as well as adjust the piece so it creates a 90-degree angle to support books.

6. Wire brush and apply a finish to the piece.

SHELF BRACKETS

SUGGESTED MATERIALS

- 3/16" x 1¼" (5 mm x 30 mm) flat stock
- Cut at 19" (480 mm) for the leaf brackets
- Cut at 15" (380 mm) for all other brackets
- 3/8" (10 mm) round stock
- 3/16" (5 mm) round stock (rivets)
- 1/8" round stock (rivets)

SUGGESTED TOOLS

- Ball-peen hammer
- Fullering tool
- Hot chisel
- Riveting tool

TECHNIQUES

- Texturing
- Riveting
- Leaves
- Curves
- Scrolls

Shelf brackets are quick projects that lend themselves to great variation. These steps are based on "general" size brackets, but they can be adapted for larger or smaller shelves as needed. The steps for each bracket L are the same, except for the leaf bracket. Variations for that bracket are highlighted throughout the steps.

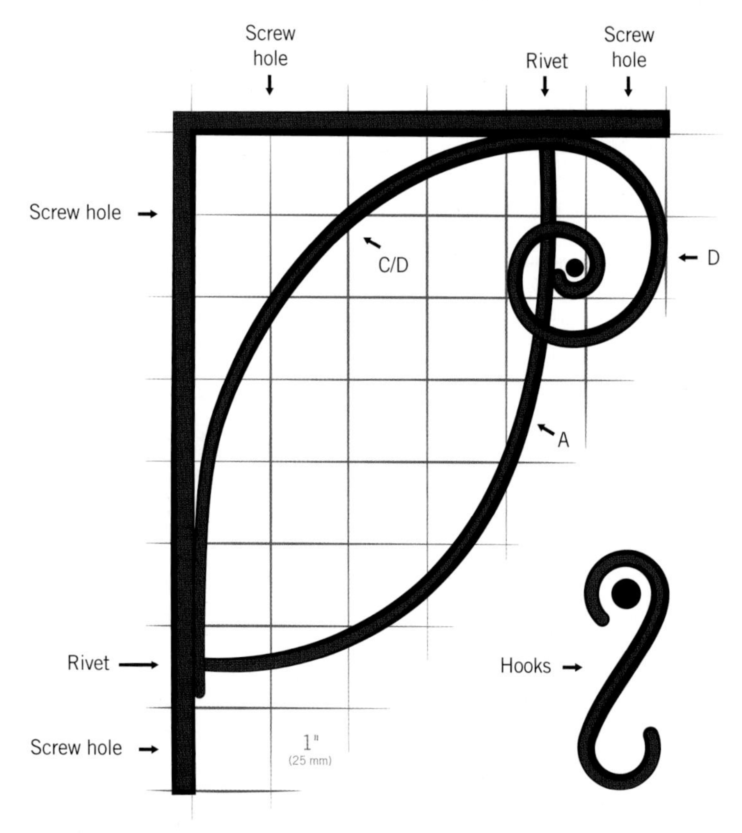

TIP: USING JIGS

When making multiple repeated shapes, a bending jig (a piece to form a shape around) can help speed up the process and increase efficiency. Make the desired piece, then make a former to match that template and you have a jig. This jig for making diagonals was welded together with scrap angle iron from an old bed frame.

BRACKET

1. Use one or more ball-peen hammers to texture the 15" (380 mm) of flat stock. Vary the location and force of your blows to create an interesting pattern (skip this for the leaf bracket).

2. Mark the flat stock 7" (178 mm) from the top end and make a tight 90-degree bend at the mark. This can be done cold with bending forks, in the vise, with a torch, or using the forge and anvil edge. If using the forge, it can help to quench the area to each side of the mark to isolate the curve before bending.

3. Cut the top of the bracket at 6" (150 mm) and the bottom at 8" (200 mm). For the leaf bracket, cut the bottom at 11" (280 mm).

4. If not making a leaf bracket, cut a 90-degree point at bottom bracket, then bevel the edges with a hammer or grinder.

5. If making a leaf bracket, forge a leaf at the bottom end with a 6" (150 mm) taper leading to the leaf.

6. Drill mounting holes and 3/16" (5 mm) holes for the rivets at the points indicated in the diagram. Use a countersink bit from the backside of the rivet holes to create a space for the flush rivets.

DIAGONALS

A. TEXTURED

Curve the 3/8" (10 mm) round stock according to the layout, running each end long. Forge a tenon on the ends down to 3/16" (5 mm) diameter and cut so approximately 3/8" (10 mm) protrudes past the brackets. Heat one end, clamp the diagonal in the vise, and rivet the diagonal to the bracket. Repeat for the other rivet. File or grind the rivet flush on the backside of the bracket.

B. LEAF

Follow steps for textured bracket. Once the piece is riveted together, get a good heat in the bottom portion of the bracket and use scrolling tongs to wrap the leaf around the diagonal.

C. BUTTONED

Curve the 3/8" (10 mm) round stock according to the layout. Cut and round the excess. Use a ball punch to forge a depression at each rivet spot. Drill or punch 3/16" (5 mm) holes and rivet together.

D. SCROLLED

Forge a taper onto one end of the 3/8" (10 mm) round stock and scroll according to the layout. Cut off the excess on the bottom end, round the end and use a ball punch to forge a depression at the bottom rivet spot. Drill or punch a 3/16" (5 mm) hole for the bottom rivet and an 3/8" (10 mm) hole for the top rivet, then rivet together. Turn the scrolled brackets into hangers as well using some of the 3/8" (10 mm) round stock as a cross piece and forging S-hooks using the 3/16" (5 mm) round stock.

MOSS BOX

BY CAITLIN MORRIS, MS. CAITLIN'S SCHOOLS OF BLACKSMITHING, FREDRICK, MARYLAND, USA

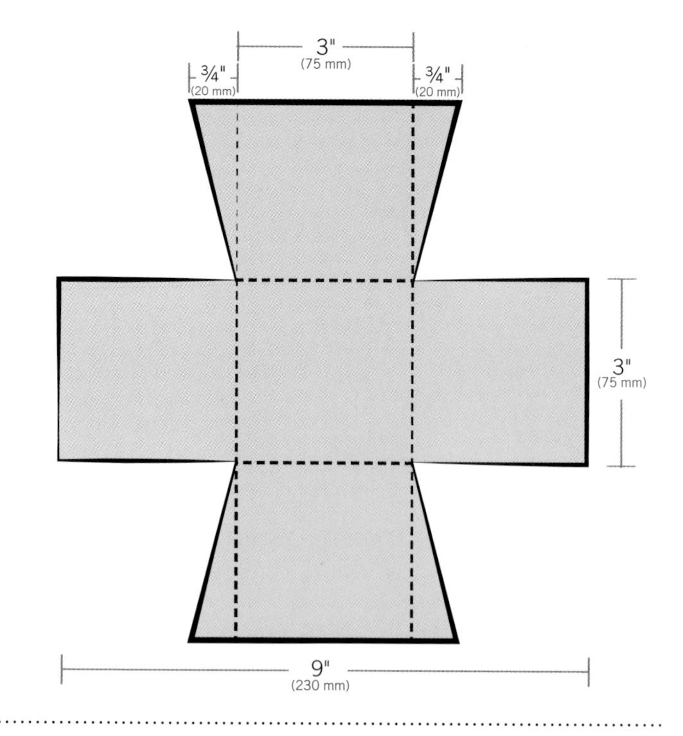

Besides being wonderfully unique, these leggy moss boxes are a great learning tool because they are a simple concept that allows for a lot of creativity and also plenty of fun challenges. The symmetry of the square box and four legs gives an opportunity to practice the same techniques, but the design does not demand precision. Moss does not need to be watered and can be obtained at your local garden supply store or online. Take the basic design and run with it, making it your own mossy creatures!

SUGGESTED MATERIALS

14-gauge sheet metal at least 10" x 10" (250 mm x 250 mm) (box)

3/8" (10 mm) square stock

4 cut to 6" (150 mm)

1/4" (6 mm) round bar (antennae optional)

2 cut to 5" (127 mm)

1/8" (3 mm) round stock (rivets)

3/8" x 3" (10 mm x 75 mm) flat stock cut to 3" (75 mm) (former)

SUGGESTED TOOLS

Rivet header

C-clamps

Dull chisel

TECHNIQUES

Cutting and folding sheet metal

Riveting

Upsetting

Tapering

BODY

1. Draw out the box shape on the 14-gauge sheet and cut the shape out with a band saw, chisel, or grinder. Clean the edges with a file or sander. Use a dull chisel to lightly work along the insides of the edges you will be bending. This weakens the metal slightly, encouraging the metal to bend along those lines.

2. If your forge is large enough, heat the entire sheet and begin making the folds using the edge of the anvil or a vise. Work the bends a little at a time and flatten out areas that get bowed as you go. If you have access to a torch, cleaner bends can be made by clamping the box to the 3/8" x 3" x 3" (10 mm x 75 mm x 75 mm) flat-sock former with C-clamps. Once the box is roughly folded, clamp the box into its final shape, heat the bends with a torch, and let it cool while clamped to make it retain the crisp box shape.

3. Mark and drill out holes for each of the rivets for the legs plus (optional) antennae.

LEGS

1. Upset each of the legs over a few heats and then use half-face blows on the far side of the anvil to isolate the foot of each leg. Rotate the metal back and forth 90 degrees every few hits (working only two of the four faces) until the "ankle" and "foot" are the desired shape and appearance. Approximately 1/4" (6 mm) square for the ankles looks nice.

2. Mark the legs at 1½" (40 mm) and 3" (75 mm) from the top (non-foot) end of the legs. The 3" (75 mm) joint will be the "knee" and the 1½" (40 mm) joint will be the "hip." Take care to orient each joint correctly for this next step (use the finished picture as a reference).

3. Heat the legs up, hold the stock at 45 degrees to the beveled edge of the anvil and hit toward the edge of the anvil with your hammer to fuller each joint down to half its thickness. Finally, flatten the last ½" (12 mm) of the leg using half-face blows to create the attachment point for the box and bend each leg at the knee and hip to finish forming the legs. Drill out rivet holes in each leg.

ANTENNAE (OPTIONAL)

1. Antennae make a fun addition to these boxes, with the attachment points serving as eyes. Forge a point into each of the 1/4" (6 mm) round stock pieces. The longer the point, the more graceful the antennae. Once the point is forged, curl each into a tight spiral, then flatten the last 1/4" (6 mm) of the uncurled end to make the attachment point. Make each antenna slightly different to add some personality to the piece.

ASSEMBLY

1. The box can be riveted together cold. Slide rivets in from the inside of the box and support the heads with horn of the anvil while using a rivet header to set the rivets.

2. If the box does not sit evenly on all four legs, rotate the tallest leg out at the rivet or hammer it lightly on the knee (cold) to unbend that joint.

3. Sand down any sharp areas, then wire brush and finish. Painting the box can be an attractive way to prevent rusting.

DISPLAY EASEL

BY BILL KIRKLEY, COLUMBIA, SOUTH CAROLINA, USA

SUGGESTED MATERIALS

- 3/8" x 1" (10 mm x 25 mm) flat stock: 3 cut to 6" (150 mm) (can be run long to begin), 1 cut to 1½" (40 mm)
- 1/8" (3 mm) round stock (rivets)

SUGGESTED TOOLS

- Hot-cutting chisel
- Needle file

TECHNIQUES

- Drawing out
- Filing
- Accurate drilling

This little easel is great for displaying photos, artwork, and holding open that cookbook (or blacksmith project book) while you follow latest recipe (or project). It is also a good example of doing a lot with limited starting materials. The entire piece is made with fewer than 20" (510 mm) of the same starting stock.

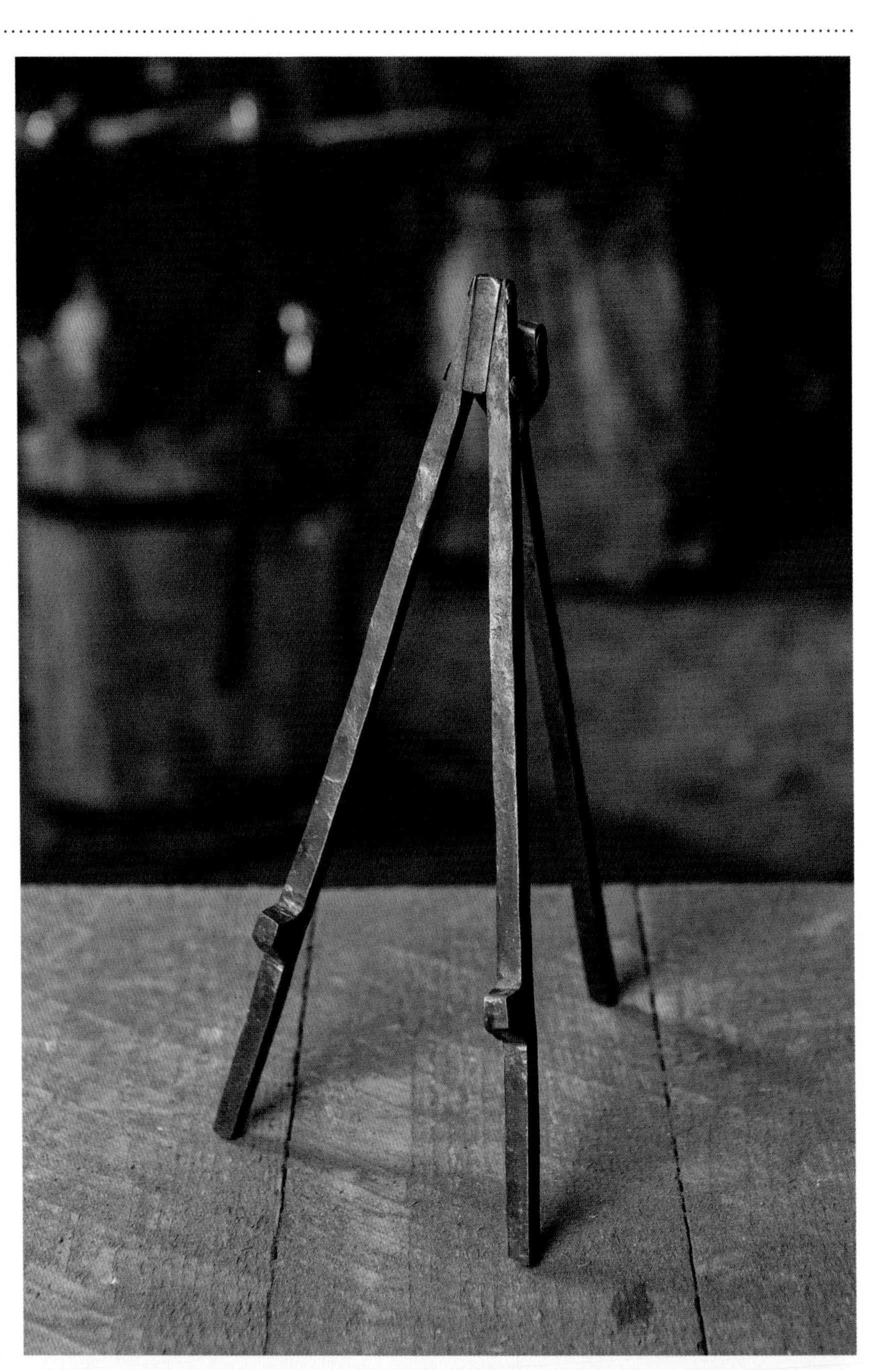

FRONT LEGS

1. Mark one of the 6" (150 mm) stocks at 1½" (40 mm) and 2" (50 mm) from one end. Set each mark using the edge of the anvil or cut-off tool.

2. Heat and place the 1½" (40 mm) set mark on the near edge of the anvil angled down about 20 degrees. Use half-face blows to form the bottom edge of the peg and begin isolating the bottom section of the leg.

3. Draw out the bottom of the leg until the cross section is ⅜" x ⅜" (10 mm x 10 mm).

4. Place the second mark on the far edge of the anvil and use half-face blows to isolate the top of the peg. Draw out the remainder of the leg so that cross-section is ⅜" x ⅜" (10 mm x 10 mm), then forge out the peg so that is perpendicular to the legs and also ⅜" x ⅜" (10 mm x 10 mm).

5. Cut the bottom of the leg 2½" to 3" (60 to 75 mm) from the bottom of the peg. Cut the peg at ¾" (20 mm) and cut the top of the leg so the total length is 12" to 13" (300 to 330 mm).

6. Position the leg so that the peg is facing straight up into the air and mark a line from the top left corner of the leg to 1½" (40 mm) down from the top right corner. Cut this line to create the surface that will attach to the hinge.

7. Repeat for the second front leg, reversing the side of the angled cut.

BACK LEG

1. Use a hacksaw or hot cut a ¾" (20 mm) slot centered on the last piece of 6" (150 mm) flat stock. Forge each arm into ⅜" x ⅜" (10 mm x 10 mm) dimensions, then draw down the rest of the stock to ⅜" x ⅜" (10 mm x 10 mm) dimensions.

2. Use a ⅜" (10 mm) spacer to align the arms parallel to each other and in line with the rest of the leg.

3. Cut down the arms to ¾" (20 mm) and round the edges with a file or grinder.

4. Using the spacer to keep the arms from bending, center punch and then drill out ⅛" (3 mm) holes centered 3/16" (5 mm) from the end of the arms.

HINGE

1. Cut the hinge out according the diagram using a grinder or hacksaw.

2. Center punch and drill out the three holes slightly larger than ⅛" (3 mm). There is low tolerance for error with this piece so pay extra attention to centering each hole. It can help to begin with a smaller drill bit and work up to the final size.

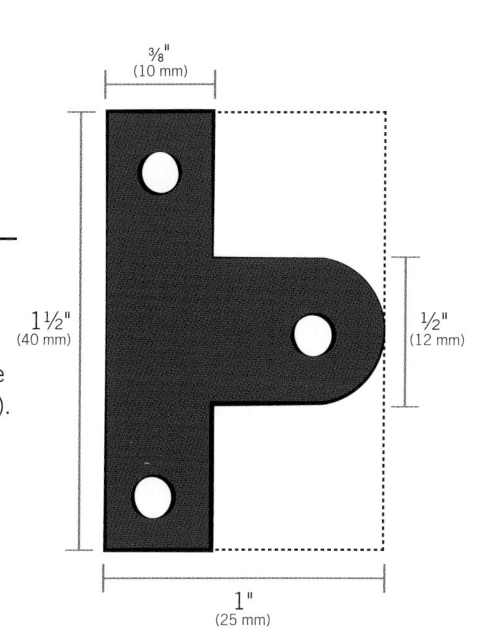

ASSEMBLY

1. Use the hinge as a template to mark the holes for the two front legs, then drill out each hole.

2. Use ⅛" (3 mm) round bar to rivet the front legs and hinge together. This can be done cold.

3. Loosely assemble the back leg onto the piece and use a ruler or straight edge to mark where the bottom of the leg is at the same length as the front legs when collapsed. Cut the back leg to that length.

4. If the arms fit loosely on the hinge, hammer the arms together slightly to create friction for the back leg and then rivet the back leg to the hinge. Then round the feet as needed to make the easel sit flat. Wire brush and finish.

CANDLE LANTERNS

BY GUNVOR ANHOJ, CALNAN & ANHOJ, WICKLOW, IRELAND

SUGGESTED MATERIALS

- ¼" (6 mm) round stock: 1 cut to 31½" (800 mm)—Spiral, 1 cut to 20" to 24" (50 to 60 mm)—Hook
- ⅛" (3 mm) round stock (ring)
- ⅝" (16 mm) round stock run long (hot cutter)
- ⅛" (3 mm) steel spacer
- Old wine bottles

SUGGESTED TOOLS

- Scrolling tongs
- 1" (25 mm) spacers

TECHNIQUES

- Curving and spiraling freehand
- Tapering
- Working with glass

This lantern project is a fun chance to combine various materials into a unique final piece as well as use the heat of the forge in a new way. Few things are cooler than getting to hammer pieces of white hot steel. But using that same piece of steel like a light saber to cut through glass bottles has to come close.

If you want to hang these lanterns from a post or wall, you can make an accompanying hanger such as the plant hanger on page 119.

SPIRAL

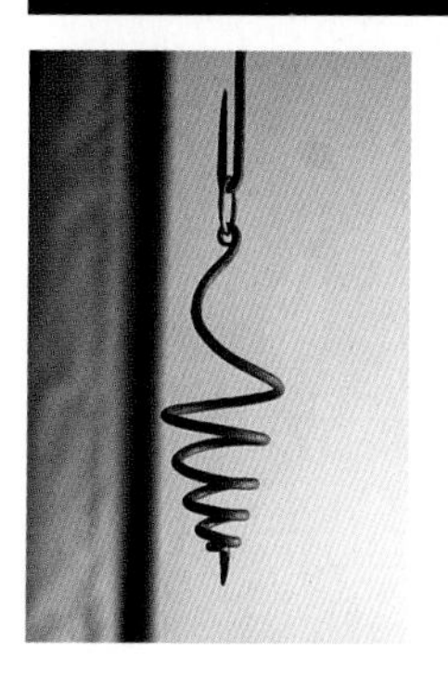

1. Forge a short taper on one end of the 31½" (800 mm) round stock, then make a right-angle bend 1" (25 mm) from the tip.

2. Get a good heat 2" to 4" (50 to 100 mm) past the bend. Clamp the tip in the vise and begin making a tight scroll. Continue until 8" (200 mm) of straight stock is left. Keep the scroll tight and even by quenching the inside after each heat.

3. Get an even heat on the scroll and clamp the 8" (20 cm) of straight stock into the vise where it meets the scroll. Use two pairs of tongs to pull the scroll into a spiral.

→

4. See where the tealight sits in the spiral and mark that spot with chalk. Use a wire, string, or chain to measure and mark the point 6¾" (170 mm) up from that point along the stock. Cut the bar at that length, taper, and curl the end.

5. Clamp the spiral into a vise and use tongs to create a gradual curve in the remaining stock that ends with the spiral centered on the piece.

HANGER AND RING

1. Forge a short, 2" to 3" (50 to 75 mm) taper on one end of the hook and then, using a piece of ⅛" (3 mm) steel as a spacer, create a tight "U" hook. This hook needs to be able to fit through the neck of the bottle.

2. Taper the other end of the hook and create a "shepherds hook" as the top of the hanger.

3. For the ring, use a piece of ⅛" (3 mm) round stock, welding rod, or something similar and form it around a 1" (25 mm) pipe or freehand on the anvil.

HOT CUTTING GLASS

The scoring created by a hot bar splits the surface of glass and eventually seperates along that score line. It is important to create as straight a score line as possible. Set up a jig to hold the bar level and at least 1" (25 mm) from the base of the bottle.

BOTTLE

1. Remove any labels from the wine bottle.

2. Get the ⅝" (16 mm) bar as hot as your forge will allow. Wire brush the bar to remove loose scale, then set on spacers and begin scoring the bottle.

→

3. Watch the score line as it develops; if there are any breaks in this line it's important to go back over these. Keep going around until the bottom comes loose with a small "tink" sound. This may take more than one heat.

4. Clean the bottle's break with fine wet sandpaper, then assemble the lanterns. Make sure to have some spare bottles, as there will be breakages while you learn the process!

PROJECTS

CHAPTER 6: KITCHEN AND BATH

There are a lot of opportunities in the kitchen and bathroom for blacksmith-made holders, racks, dispensers, and the like. These are also usual locations for sets of fixtures to share common design motifs. In the bathroom fixtures for this chapter, particularly, a unique style is highlighted for each project, but steps for carrying each style over to the other fixtures is also discussed. As you move through this chapter's projects, it is a good point in the text to start thinking about—and maybe even incorporating—your own design elements and variations into project steps.

Pieces going into the kitchen and bathroom will be subjected to greater moisture levels than other places around the house. Consider finishes with greater water resistance for such pieces, or cover finished pieces with a good-quality clear coat for additional moisture protection.

NAPKIN RINGS

SUGGESTED MATERIALS

- $^{3}/_{16}$" (5 mm) round stock, various lengths

SUGGESTED TOOLS

- 1½" (40 mm) pipe or round stock
- Small scrolling tongs
- Pliers

TECHNIQUES

- Tapering
- Wrapping
- Texturing
- Curling

Napkin rings are a quick and playful form that spice up any table setting. Try some of these styles, and then branch out into some of your own design. Make sure to thoroughly wipe down the rings after finishing so they don't dirty up your good napkins.

SCROLLED ENDS

1. Start with a 13" (330 mm) length of round stock. Taper and curl each end into ½" (12 mm) loops in opposite directions.

2. Secure the pipe vertically in a vise. Get an even heat along the entire stock and wrap the piece around the pipe. Use pliers or tongs to hold one end against the pipe and another pair of pliers or small tongs to do the wrapping. Clean the piece thoroughly and finish with a food-safe finish.

→

VARIATIONS

A.
Start with 13" (330 mm). Taper and scroll one end tightly several times onto itself. Taper the other end and use pliers to create a squiggle in the opposite direction as the scroll. Make the total length 12" (300 mm). Then follow step 2.

B.
Start with 14" (355 mm). Forge a leaf on one end and a round taper on the other. Use pliers or small scrolling tongs to forge some squiggles in opposite directions on each end. Make the total length 12" (300 mm). Then follow step 2.

C.
Start with 12" (300 mm). Flatten and scroll one end, then flatten the rest of the piece and scroll the other end for a total length of 10" (250 mm). The follow step 2.

D.
Start with 10½" (265 mm). Flatten the full length and texture with the ball peen. Round the ends with a file or grinder. Then follow step 2.

E.
Start with 10" (250 mm). Flatten and taper the piece into an elongated pizza shape approximately 11" (280 mm) long. Put slight crooked scrolls on each end. Then follow step 2.

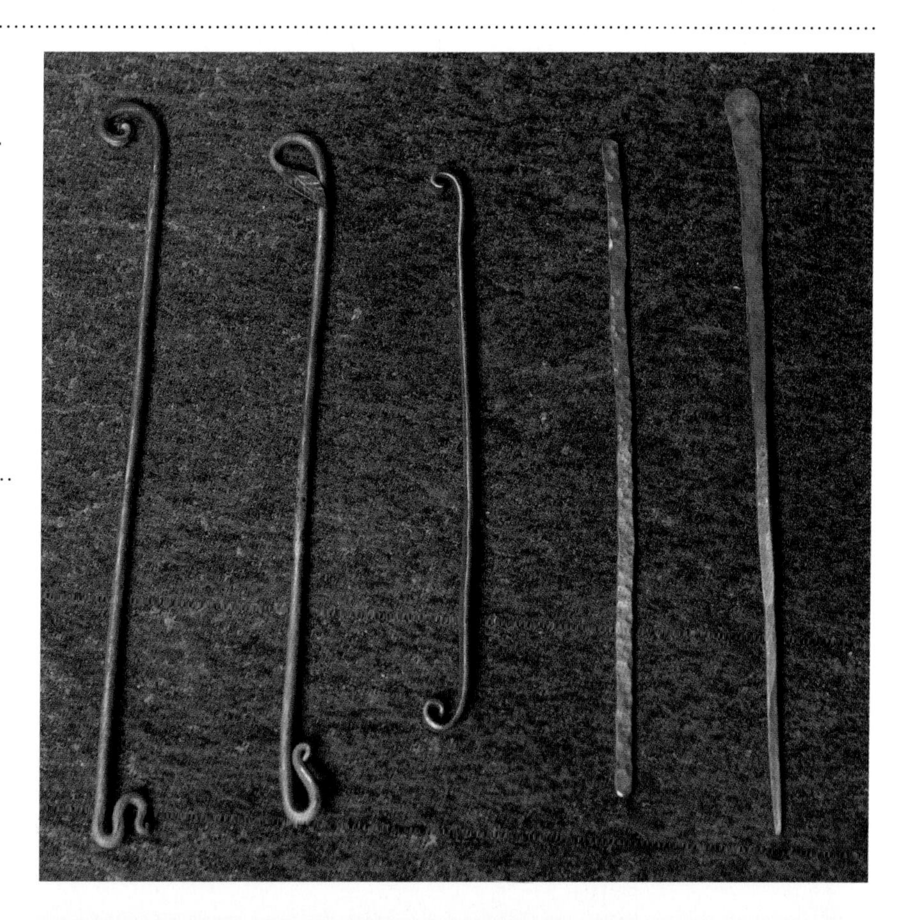

POT RACK

SUGGESTED MATERIALS

Pot Rack		Key Rack	
¼" x 3" (6 mm x 75 mm) flat stock: 1 cut to 32" (812 mm)	½" (12 mm) square stock	¼" x 3" (6 mm x 75 mm) flat stock: 1 cut to 12" (300 mm)	½" (12 mm) square stock: 2 cut to 3" (75 mm), 2 cut to 2" (50 mm)
¼" x 1" (6 mm x 25 mm) flat stock: 1 cut to 13½" (343 mm), 1 cut to 21½" (546 mm)	2 cut to 5" (127 mm)	¼" x 1" (6 mm x 25 mm) flat stock: 1 cut to 4" (102 mm), 1 cut to 7" (178 mm)	¼" (6 mm) round stock
	2 cut to 4" (102 mm)		
	¼" (6 mm) round stock		

Whether they're for a pot, pan, cup, coat, key, leash, or some variation of the above, these racks should fit the bill. Both racks are set up with hangers at two distances from the wall to provide users with extra flexibility depending on what they wish to hang.

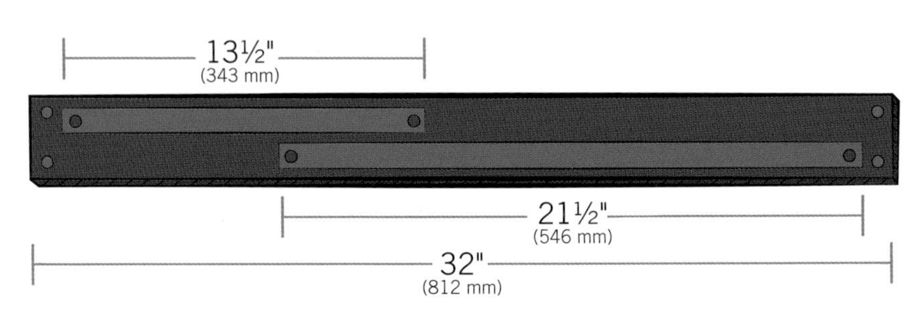

SUGGESTED TOOLS

Riveting tool

Monkey tool for 3/8" (10 mm)

TECHNIQUES

Riveting

Tenons

Texturing

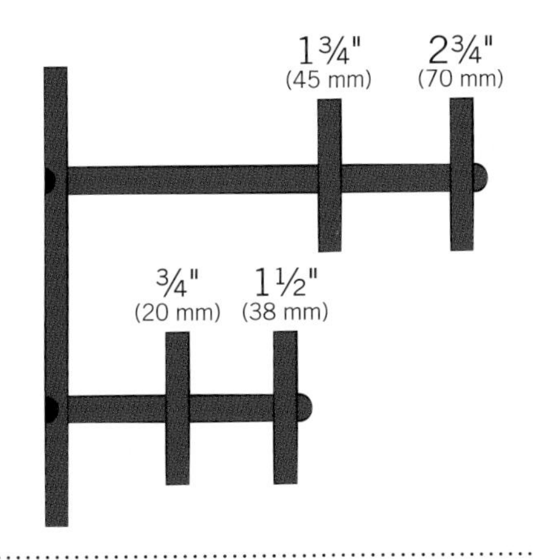

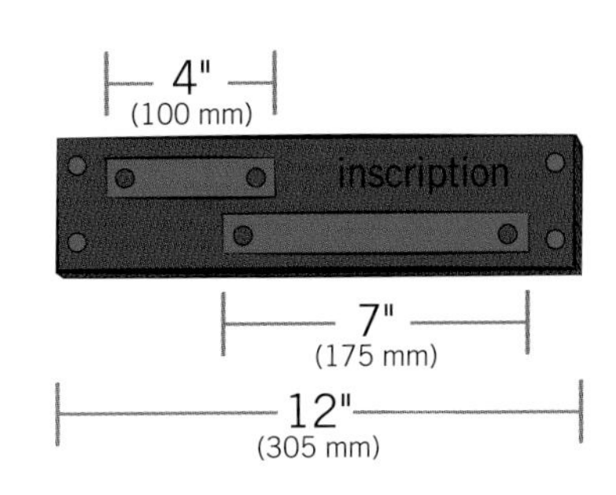

1. Forge slight tapered bevels into each corner of the mounting plate using angled hammer blows.

2. Forge a textured bevel around the outside edge of the mounting piece using a large round-faced hammer.

←

3. Drill or punch 3/16" (5 mm) mounting holes and 3/8" (10 mm) rivet holes. Use a countersink bit to make a large bevel on the backside of the mortise holes. Drill out matching mortise holes in the 1" (25 mm) flat-stock racks.

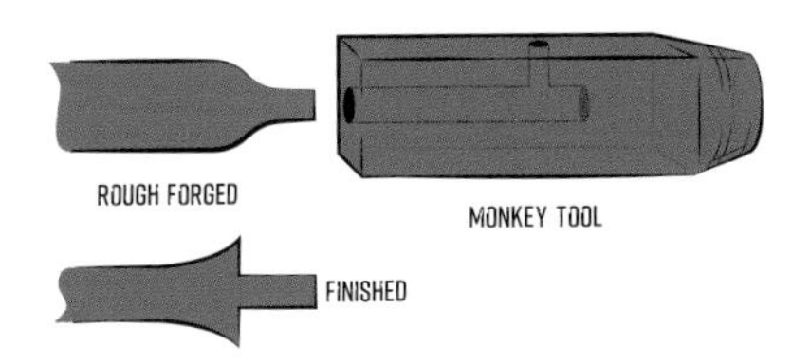

NEW TECHNIQUE: TENONS

Tenons fit into mortises (holes) and together make up a traditional joinery unit. There are a number of special tools such as a spring swage or guillotine tool that can assist the creation of the tenon.

A simple way without such tools is to use the near edge of the anvil and carful full-face blows to isolate and then form the tenon. Tenons are finished with a "monkey tool," a piece of solid stock with a hole drilled in to upset the parent stock slightly to square off the shoulder for a tight fit.

SPACERS

Forge 3/8" (10 mm) round tenons on each end of the square stock spacers. For a pot rack, make two spacers with 2¾" (70 mm) between tenons and 1½" (40 mm) between the other two. For a key rack, make two spacers with 1¾" (45 mm) between tenons and ¾" (20 mm) between the other two.

HOOKS

Forge a tight taper in a length of ¼" (6 mm) round stock and curl into the tip of the hook. Quench the tip and make another larger curve about 1" (25 mm) in diameter to complete the hook shape. Cut off approximately 2½" (65 mm) as measured from the tip of the hook. Square off 1" (25 mm) and fold that section back into a tight "U" shape, using a piece of ¼" (6 mm) flat stock as a spacer (see Twisting Forks project, for example).

ASSEMBLY

Rivet the racks to the mounting bracket and grind the rivets flush on the backside. Wire brush the piece and finish.

TOILET PAPER HOLDER

BY WILLIAM PINDER, W. A. PINDER IRONWORKS, DORCHESTER, ENGLAND, UK

SUGGESTED MATERIALS

- ¼" x ¾" (6 mm x 20 mm) cut to 12" (300 mm)

SUGGESTED TOOLS

- Small cross-peen hammer
- Hot-cutting chisel
- Scrolling forks

TECHNIQUES

- Leaves
- Hot cutting
- Tapering
- Punching

Each of the next few bathroom fixtures highlights a different design style, with notes for making matching sets across the fixtures appearing at the end of each project. Most functional projects provide certain general design constraints—how tall, how far off the wall, how wide, and so on. To these constraints, the artist and the craftsperson adds his or her own design style. These alternatives give an example of that process as well as the way in which motifs can be carried across projects. This series of projects also provides an opportunity to practice incorporating your own designs across multiple pieces.

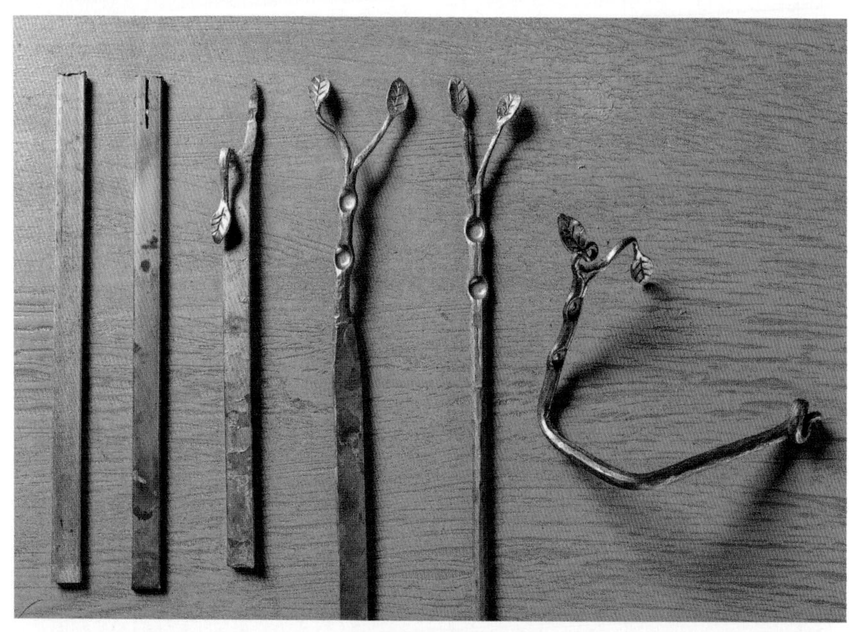

1. Hot cut an approximately 1½" (40 mm) line down the middle of one end.

2. Draw out the two ends to approximately ¼" (6 mm) square.

3. Forge a leaf into each end and forge a round taper into the stems.

4. Draw out the section below the split to approximately ⅜" (10 mm) square and 3" (75 mm) long and then round the section.

5. Use a ball-end punch to make two indents for screw holes.

6. Draw out the remaining stock to approximately ⅜" (10 mm) square and 13" (33 cm) long. Cut off any excess, then forge a round taper at the end.

7. Shape this portion into the final holder either free hand or by clamping the piece into a vise and using tongs and scrolling forks. The holder should be approximately 3½" (90 mm) off the wall with a 5" (127 mm) bridge for the roll to sit on.

8. Make a small ridge at the tapered end to keep the roll in place.

9. Wire brush and finish with a water-resistant finish.

ROPEY

Starting stock: ⅝" (16 mm) round stock cut to 8" (200 mm)

1. Leaving 3½" (90 mm) on one end, forge a round taper into the stock for a total length of 18" (455 mm).

2. Upset the thicker end slightly and then offset the upset with a few angled blows.

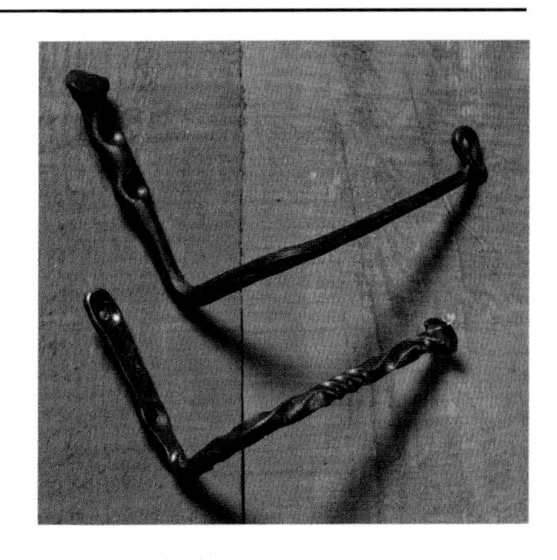

3. Use a ball punch to make indents for screws holes at approximately 1½" (40 mm) and 3" (75 mm) from the upset.

4. Forge a small curl on the tip, then shape into the final form following step 7 above.

REVERSE TWIST

Starting stock: ⅜" (10 mm) square stock cut to 13" (330 mm)

1. Create a reverse twist pattern along 9" (225 mm) of stock, leaving 1" (25 mm) untwisted at one end.

2. Upset the untwisted end over several heats to create a stopper for the toilet roll. If upsetting in a vise, quench the twisted portion before each upset to avoid marking the twists.

3. On the other end, flatten the last 3" (75 mm) of stock for mounting on the wall.

4. Use scrolling forks to make two tight right-angle bends and shape the piece into its final form. Make the piece sit about 3½" (90 mm) off the wall and 5" (127 mm) across.

TOWEL RACK

SUGGESTED MATERIALS

⅝" (16 mm) round stock cut to 32" (815 mm)

SUGGESTED TOOLS

Ball punch

TECHNIQUES

Tapering

Curling

Upsetting

Punching

This ropey design will work your arms and test your finesse. The work stems from the long taper, while the finesse comes from making the graceful curves that define the piece.

1. Upset one end of the stock and then purposely offset the upset by about 45 degrees.

←

2. Forge a 16" (400 mm) round taper into the piece, for a total length of 48" (1220 mm). Leave the first 4" to 5" (102 to 127 mm) the original thickness.

3. Working backward from the tip, curl the end over itself twice. This can be done freehand over the anvil or by using scrolling tongs. Use approximately 10" to 12" (25 to 300 mm) of stock for this portion.

4. Use a ball punch to make an indent centered on the larger curve for a screw hole.

5. Create a gradual swooping curve along the center length of the rack. This can be done cold with scrolling forks moving the piece after each bend. The rack should be approximately 28" (710 mm) wide.

6. Heat the piece again to make the last curve back to the upset end. Use a ball punch to make two indents about 1½" (40 mm) and 3" (75 mm) from the upset.

7. Make any needed adjustments so the piece sits flat and about 3" to 4" (75 to 102 mm) from the wall. Wire brush and apply a water-resistant finish.

LEAF

1. Start with ½" (12 mm) round stock cut to 48" (1220 mm). Make marks 3" (75 mm), 6" (150 mm), and 9" (225 mm) in from one end. Forge a leaf and taper the stem back to the original 3" (75 mm) mark, drawing the stem to about 5" (127 mm). Repeat for the other end.

2. With scrolling forks or over the edge of the anvil, make bends at the original 6" (150 mm) and 9" (225 mm) mark on each end. It is easier to make the bend at the 6" (150 mm) mark first. Use scrolling tongs to shape the ivy. Use a ball punch to make two indents for screws centered just inside the original 3" (75 mm) and 6" (150 mm) marks.

REVERSE TWIST

1. Start with ½" (12 mm) square stock cut to 40" (1015 mm). Make marks 3" (75 mm) and 6" (150 mm) from each end.

2. Create a reverse twist pattern along the center section. Bend into the rack shape with 90-degree bends at the original two marks.

3. Forge a shoulder at the final 2" (50 mm) of each end and square off the shoulder. This last step can be done before the bends, but if doing so, be sure to make the shoulder on the correct face of the stock to account for the two bends.

HAND TOWEL RING

SUGGESTED MATERIALS

- 3/8" (10 mm) square stock cut to 19" (480 mm)
- 3/8" (10 mm) square stock cut to 6" (150 mm)

SUGGESTED TOOLS

- Twisting wrench
- Rawhide mallet
- Ball punch

TECHNIQUES

- Tapering
- Twisting
- Punching

At some point, machines were created that could produce metal components such as twisted pickets (the vertical pieces) in gates, rails, and fencing with great uniformity and speed. Many modern gates with those perfect twists are the product of such machines and not a blacksmith's hand. That's why I like funky reverse twist patterns so much. It's the intentional irregularity that shows that a human, not a machine, made this piece.

RING

1. Forge a reverse twist pattern along the center 22" (560 mm) of the square stock. Create variations by adjusting the direction of twists, the length between reversals, and the number of twits each heat. Before each heat, quench the prior twists to prevent the piece from untwisting or being marked by the vise.
→

2. Taper each end to a point. Forge a 1" (25 mm) diameter loop on one end and a tight curl in the same direction on the opposite end and then bend the piece into a loop. If making the loop on the anvil, use a rawhide mallet if you have one to avoid denting the twists.

BRACKET

1. Forge a 3" (75 mm) round taper on one end of the ⅜" (10 mm) square stock. Begin to curl the taper into a 1" (25 mm) loop, leaving enough space to attach the ring.

2. On the other end, use a ball punch to create two indents for screws along the 2" (50 mm) of the stock.

3. Make a 90-degree bend just past the second indent. Heat the partially curled end, attach the ring, and close the loop.

4. Drill out the holes for mounting. Wire brush and apply a water-resistant finish.

ROPEY

Start with a 16" (405 mm) piece of ⅝" (16 mm) round stock and taper to 24" (610 mm), leaving the first 2½" (65 mm) untapered. Upset that end and then offset the upset with angled blows. Use a ball punch to make two indents for screws next to the offset and then curve the piece into the final form. Work back from the tip, starting with a small curl at the end.

LEAF

Forge the hand towel ring in the same manner as the toilet paper holder, but start with an extra 4" (100 mm) of stock to account for the ring size (approximately 16" [400 mm] of starting stock). When making the ring, start by folding over and wrapping the tip around the parent stock and then form the rest of the ring with forks or over the horn of the anvil.

PAPER TOWEL HOLDER

SUGGESTED MATERIALS	SUGGESTED TOOLS	TECHNIQUES
Base: ¼" (6 mm) steel plate, at least 8" x 8" (200 mm x 200 mm)	Monkey tool	Riveting
Center rod: ⅝" (16 mm) round bar cut to 16" (400 mm)	Riveting tool	Tenons
Scrolls: ⅜" (10 mm) round bar: 1 cut to 19" (485 mm), 1 cut to 22" (560 mm)	Medium bending forks	Scrolls
Rivets: ¼" (6 mm) round bar		Wrapping
14-gauge copper wire		

Copper contrasts beautifully with steel and makes a great accent to projects. In addition to the copper wraps, this is one of the more complicated projects in the book in terms of the combination of materials, techniques, and assembly. Be sure you are very comfortable with riveting and scrolling before taking this one on.

1 box = 1" (25 mm)

BASE

1. Cut out a 7½" (190 mm) circle in the steel plate using an angle grinder, torch, plasma cutter, or vertical bandsaw.

2. Drill a ⅜" (10 mm) hole in the center of the circle and countersink one side with a wide countersink bit. This is where the center shaft will be riveted.

SCROLLS

1. Draw the scrolls using chalk beforehand for reference during the forging process.

2. Heat up and flatten the lengths of ⅜" (10 mm) round stock until they have approximately ½" (12 mm) of a non-beveled flat surface. Taper the ends 3" (75 mm).

3. Optional: For some extra design interest and to create a surface for personalization (letter stamping), flatten about 5" (125 mm) of the taper with angled hammer blows to create a blade-like taper. Make the thinner edge the outside of the curve when scrolling.

4. Forge the first 1½" to 2" (40 to 50 mm) curls of each scroll on the anvil. Quench and repeat for the other ends. For the S-scroll (the 22" [560 mm] bar), create the second set of curls facing the opposite direction as the first set. For the C-scroll (the 19" [485 mm] stock), do second curls in the same direction.

5. Use bending forks to finish curling the scrolls according to the layout. This can be done cold or hot.

6. Get an even heat on the scrolls, hold them lengthwise on the horn of the anvil (one at a time), and use gentle hammer strokes to curve them along their width until they match the curve of the circular base. You can also use a swage block for this step.

CENTER SHAFT

1. Forge a ⅜" (10 mm) tenon on one end of the shaft and cut down the tenon so that ¼" (7 mm) protrudes when placed in the baseplate.

2. Cut the shaft down to 12½" (320 mm) from the start of the tenon. Upset the end without the tenon slightly and drill a 5/64" (2 mm) hole ½" (12 mm) into the shaft ½" (12 mm) from the top, and another 1½" (40 mm) from the top.

ASSEMBLY AND FINISHING

1. Heat the tenon in the forge or with a torch and rivet it into the countersunk hole.

2. Make sure the scrolls fit together and match the curve of the base plate, making any adjustments as necessary with bending forks. Mark where each scroll meets the base plate, both on the scrolls and the base plate. Center punch and drill out each spot with a bit slightly larger than ¼" (6 mm). Countersink the holes in the underside of the base plate with a wide countersink bit.

3. Rivet the scrolls to the base plate from the underside of the piece, supporting the bottom of the rivet with the horn of the anvil. Make the rivet flush on the underside, grinding any excess.

4. Wire brush the piece.

5. Wrap the spot where the scrolls meet with the copper wire. Begin the wrap with the wire in an L shape so that the beginning of the wrap can be covered over itself in successive wraps. When finishing the wrap, cut the wire, curve the end, and then tuck the end back into the wrap to keep the end hidden. It is best to do this on the backside of the V gaps where the two scrolls separate.

6. Wrap the top of the shaft with the copper, starting one end in the top hole and finishing the other end by feeding it into in the bottom hole.

←

7. Lightly wire brush the copper and finish the piece. Add felt pads to the base to prevent scratching.

PROJECTS

CHAPTER 6: COOKING

Food and drink enthusiasts have a growing interest in not only what we cook but in how we do so and in the cultural significance of our cooking choices. Demand flows naturally from these enthusiasts to the blacksmith's craft. Loggerheads, once made extinct by the microwave, have come back to spice up the holiday cocktail party. Railroad spikes, once used on—you guessed it—railroads, are now forged into everything from bottle openers to oyster shuckers. And not long ago, a blacksmith-made egg spoon became the center of an ongoing cultural cooking war centered around what and how we as a society should be eating. This chapter contains a mix of traditional and modern projects that represent these renewed interests and shifting attitudes.

Cooking utensils require particular protection from moisture similar to the projects in the last chapter. However, finishes used on these projects also have to be food safe. Lightly oiling pieces before and after use, just like a cast iron pan, is a simple, food-safe solution to this moisture dilemma. Heat the piece slightly first to help the oil penetrate better and wipe away excess with a paper towel or rag. Another important design consideration is minimizing surfaces on which bacteria can develop. Harmful bacteria won't develop on steel, but they will develop on bits of food lodged in small crevices within a forged piece. To prevent this, heavily burnish and sand smooth any surfaces that will be in contact with food before applying a finish.

LOGGERHEAD

BY KERRY RHODES, FORGED CREATIONS, DELAWARE CITY, DELAWARE, USA

SUGGESTED MATERIALS

¾" (20 mm) round bar

SUGGESTED TOOLS

Fullering tool

TECHNIQUES

Drawing out

Handles

Loggerheads are an old-school tool for quickly heating drinks before modern conveniences such as microwaves existed. Use them the same way you quench a piece of stock from the forge. The unique way they sizzle drinks also creates distinct flavor profiles by, for instance, caramelizing sugars. These are a great seasonal project for holiday gifts or sales, especially for drink enthusiasts. In lieu of an open fire (not as common in kitchens these days) a gas burner or propane plumber's torch can also be used to heat the loggerhead.

IN THE SHOP WITH FORGED CREATIONS

Just like blacksmiths of previous generations, Kerry Rhodes has become engrained in his community over the nearly 30 years he has been running his shop. One of the ways he builds that sense of community is with annual holiday parties hosted at his forge where folks get a chance see what a modern blacksmith shop looks like up close. One of the biggest hits at these get-togethers is—you guessed it—loggerhead-made hot drinks.

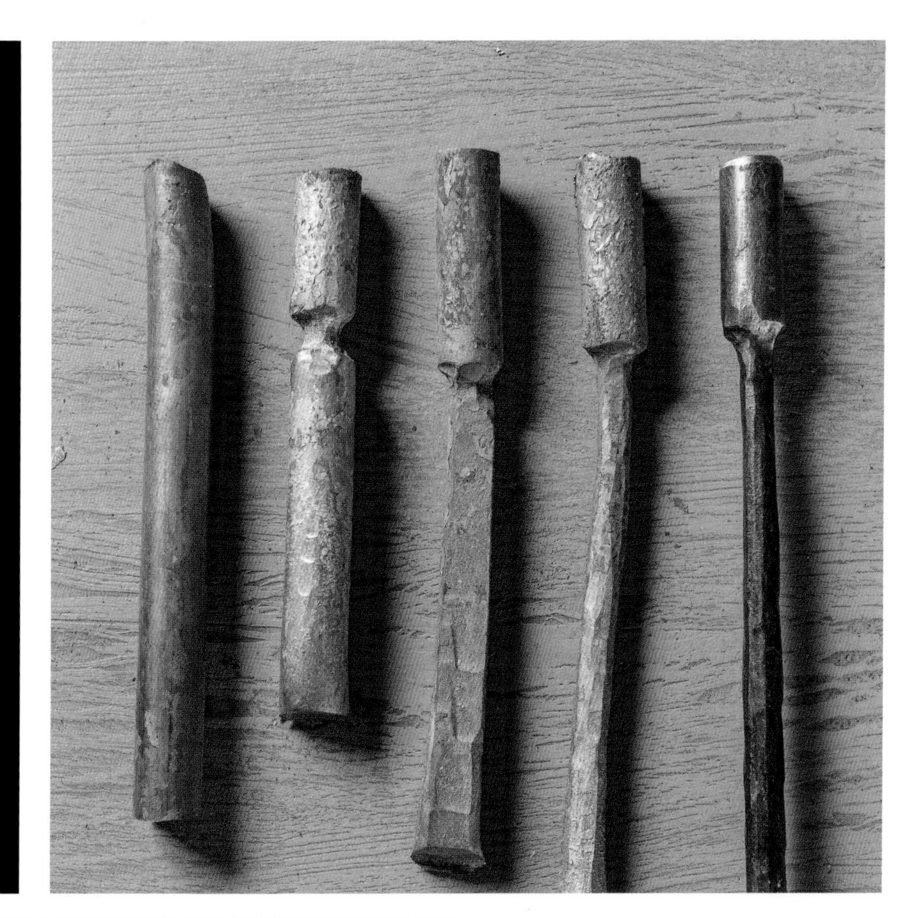

1. Neck in the stock 2" to 3" (50 to 75 mm) from one end.

2. Draw down the shaft to ½" (12 mm) round and approximately 14" (355 mm) long.

3. Cut the excess stock, leaving room to forge your favorite handle.

4. Wire brush thoroughly and finish with a food-safe finish.

LOGGERHEAD SPECIAL

Imagine it's a crisp winter evening in 1850 and you've just taken a seat at your local ale house after a long day in the smithy. With a nod, the barkeep begins making your favorite cold weather drink, "the Flip." The barkeep mixes beer, spirits, and sugar, takes a hot loggerhead out of the fire (one you made, of course) and immerses it into his concoction. The sugar caramelizes instantly and the heat creates a heavenly froth that spills over the sides of the mug and onto the counter. Satisfied, the barkeep hands you your drink and returns the loggerhead to the fire, where it gets warmed up for the second round.

RAILROAD SPIKE BOTTLE OPENERS

SUGGESTED MATERIALS

- Railroad spikes

SUGGESTED TOOLS

- 1" (25 mm) square bolt tongs
- Large twisting wrench
- Slot punch and round drift
- Cutting chisel

TECHNIQUES

- Twisting
- Punching and drifting
- Hot cutting or grinding

As an abundant and versatile starting material, many blacksmiths make and sell railroad spike this-or-thats. For the same reasons, railroad spike products can easily seem overdone. A good way to differentiate your spikes, while experimenting with new processes, is to incorporate twist patterns. The twists covered in this project can be applied to any projects using square stock. Bolt-style tongs help tremendously for handling the spikes while forging.

LOOP OPENER

1. Before forging, grind off and round the tip of the opener

2. Heat and flatten the first 1" (25 mm) of the spike to 3⁄16" (5 mm) using half-face blows. Orient the lip that sticks out at the head of the spike so that the shoulder created on the tip is on the opposite side of the opener. This makes the final opener more ergonomic.

3. Use a slot punch and drift to create an approximately 5⁄8" (16 mm) hole centered at the tip.

4. Using the horn of the anvil, enlarge the loop to 1" to 1¼" (25 to 30 mm) in diameter. Shift the loop after each hit so as to keep the loop even and flip the piece around every several hits to keep the hole from beveling in one direction.

→

5. With the loop at an even heat, hit back toward the spike to create the oval shape of the opener. If the final loop isn't symmetrical, it can be opened up on the horn again and reformed.

6. Forge the tab that will catch a bottle cap with a ball punch.

HANDLE OPTIONS

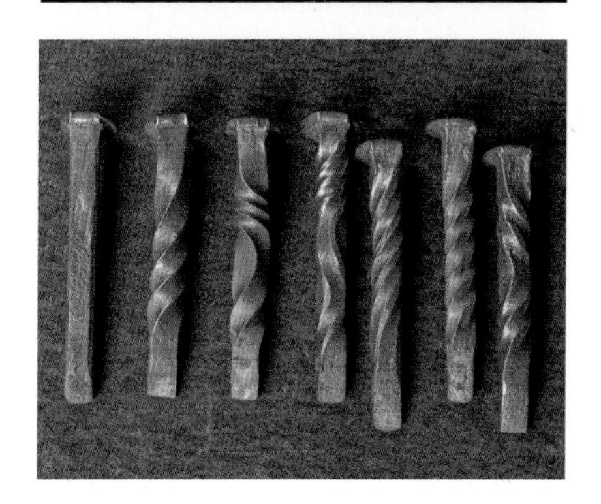

A. STRAIGHT HANDLE

Bevel the edges slightly and you're done. While simple, a straight handle shows off the nice surface texture of aged railroad spikes.

B. NORMAL TWIST

Get an even heat along the hole length of the spike. Clamp the loop end into a vise just past the shoulder and use a twisting. wrench to twist the opener 360 degrees.

C. REVERSE TWIST

Follow step B, then reheat the opener. Take out the opener and quench the first ½ of the spike, clamp the quenched portion in the vise, and turn in the reverse direction.

D. BEVELED TWISTS.

Before doing steps B or C, bevel the edges of the spike.

E. FLATTENED TWISTS

After step B or C, hammer the piece back to square.

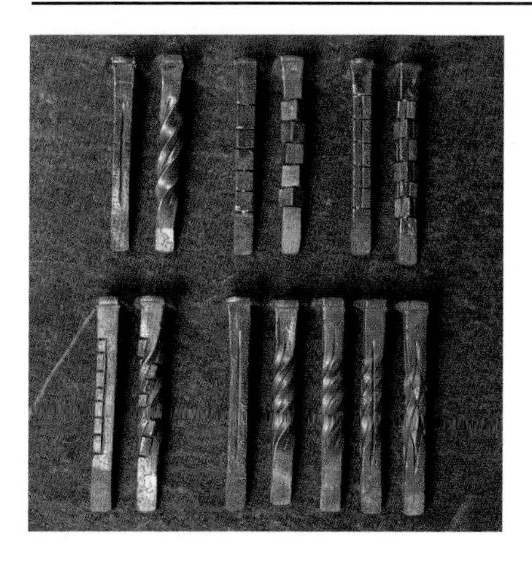

NOTE:

Any of the following can be made using a hot chisel or grinding cutoff wheel. I like to start all lines 1" (25 mm) from the head of the spike and ½" (12 mm) from the loop of the opener.

A. GROOVED LINES

Chisel or grind lines along opposing or all four lengths, then follow steps normal or reversed twists.

B. SQUARE TWISTS

Chisel or grind lines around the opener at regular or irregular intervals (⅝" or 16 mm looks nice on a railroad spike), then move onto steps B or C.

C. ALL SQUARES AND ALL LINES

Combine steps F and G and then follow step B or C.

D. ALL LINES AND PARTIAL SQUARES (STAIRCASE)

Follow step F, then create lines at regular intervals, connecting two sets of existing lines along oppose lengths of the opener. Then follow step B.

E. PINEAPPLE TWISTS

Follow steps for the grooved lines, then do a flattened twist. Add grooved lines again and then get an even heat along the length and twist in the opposite direction of the original twist.

HERB CHOPPER

BY CHRIS MOORE, ARTISTIC IRONWORK
ORATIA, AUCKLAND, NEW ZEALAND

SUGGESTED MATERIALS

$^{3}/_{16}$" x 1½" (5 mm x 40 mm) flat stock cut to 10⅝" (270 mm)

SUGGESTED TOOLS

- File or grinder
- Medium flat stock tongs
- Small round stock tongs

TECHNIQUES

- Offsetting
- Tapering
- Curling
- Beveling

This nifty herb chopper can be used one- or two-handed to cut up fresh herbs and spices. The elegance of its design comes from the symmetry of the blade and handles; take care when forging to make the tapered ends of the handles match as well as the end curls. Those are two steps that are very difficult to adjust after the fact. As with other cooking tools, stainless steel is an optional starting material to make the final chopper more resistant to rust.

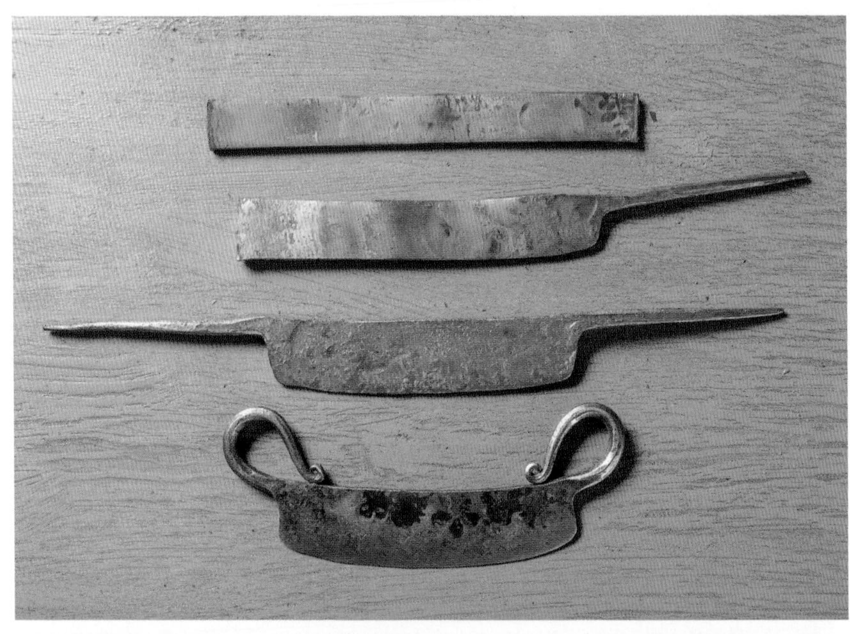

1. Offset one end of the flat stock at 2" (50 mm) using half-face blows and forge a round taper to approximately 5½" (140 mm).

2. Repeat for the other end. If using a propane forge, you may want to wait until just before curling the handles to finish tapering the piece to length so as to minimize scale buildup and stresses on the tip.

3. Forge a bevel on the edge of the remaining flat stock to create the blade. Work both sides of the blade and take care to forge the blade evenly along the piece, working back and forth a bit at a time, rather than heavily at each spot. This will ensure a symmetrical blade as well as a symmetrical curve to the chopper from the spread steel.

↗

4. Create each handle by first curling the tip in the same direction as the blade and then creating a larger curve back onto the top of the chopper. Mark the bends as a reference for each handle, or try to do each side freehand.

5. Use a grinder or file to sharpen the blade to the desired sharpness and fix any asymmetry in the blade. Wire brush and finish the piece with food-safe finish. Oil lightly after use to prevent rust.

EGG SPOON AND LADLE

SUGGESTED MATERIALS

- 16-gauge sheet metal at least 6" x 6" (150 mm x 150 mm)
- 3/8" (10 mm) round stock run long
- 10-gauge copper wire

SUGGESTED TOOLS

- Small flat tongs
- Compass
- Cold-cutting chisel
- Small punch and drift

TECHNIQUES

- Fullering
- Dishing
- Drawing out stock
- Punching and drifting
- Riveting

These artisanal accessories will spice up any cooking space, be it a Manhattan apartment or a colonial reenactment site (did they use egg spoons back then?). A note on handles: Each cooking accessory in this book uses a different handle for variety, but any one handle can be used across projects to make a matching set.

BOWL

1. For the egg spoon, draw a 6" (15 cm) circle in the sheet metal and cut out the circle using a cold-cutting chisel or grinder. For the ladle, make a 4" to 5" (100 to 125 mm) circle. Clean up the edges with a grinder or file.

2. Dish the piece using a rounded hammer face or ball peen and a former such as a swage block. For the ladle, make the bowl deeper than for the egg spoon. A simple dishing former can be made by creating a 3" (75 mm) loop from ½" (12 mm) round bar. Make the loop at a 90-degree angle from the rest of the round bar and cut off a 2" (50 mm) section of the excess. Flatten that section slightly and clamp it in the vise.
←

HANDLE

1. Round one end of the round bar and use a fullering tool to isolate ¾" (20 mm) of the end of the bar. Taper the neck back 3" (75 mm).

2. Use a cross-peen hammer to spread the ends into an oval shape, with the wide section perpendicular to the shaft of the handle. Curve the oval end to fit the curve of the bowl.
→

3. For the egg spoon, forge a gradual 16" (400 mm) taper leading to the spread end. For the ladle, forge an 11" (280 mm) taper. Leave enough material at the tip to support the bowl end.

4. Flatten 6" (150 mm) beyond the taper and cut off at that point. Use a slot punch and ½" (12 mm) drift to make a hole at the top of the handle. Clean the hole on the horn of the anvil.

5. Bevel the edges of the handle so that a cross-section of the handle looks like a compressed pyramid.
→

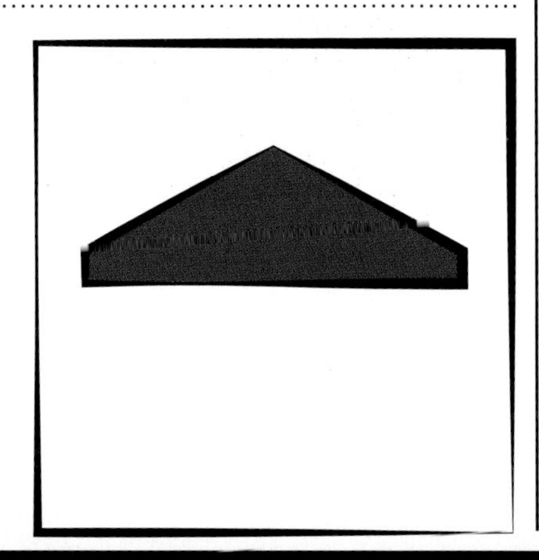

ASSEMBLY AND FINISHING

1. Drill out two holes slightly larger than 10-gauge wire in the bowl and handle. Use the 10-gauge copper for the rivets.

2. Curve the handle, resand the bowl face to a smooth polish, and coat the piece in oil. For the egg spoon, the handle should be curved so that the bowl is level when the handle is held at approximately 30 degrees. For the ladle, the handle should be curved so that the bowl is level when the handle is held at approximately 80 degrees.

COLONIAL KEYHOLE SPATULA

BY NICHOLAS K. DOWNING, DOWNING ARTS
PORTLAND, MAINE, USA

SUGGESTED MATERIALS

¼" x 1" (6 mm x 25 mm) cut to 7" (178 mm)

SUGGESTED TOOLS

Flat stock tongs

File

TECHNIQUES

Spreading

Necking

This traditional spatula design is a great way to exercise diverse hammer techniques as well as your sense of proportion. Once you get comfortable with the design, see how many of the steps you can do by eye instead of measuring.

IN THE SHOP WITH DOWNING ARTS

Nicholas Downing is a metalsmith who loves working on diverse projects, from steeled tools and colonial metalwork to hand-forged knives and custom fine jewelry. Splitting his time between a forge and a jeweler's bench is one of the ways he's developed an eye for detail in his hand-forged work. His advice to new smiths aiming for similar levels of precision: slow down and be patient. Work deliberately and don't be afraid to start over. You will be surprised how much better your second (or sixth) attempt turns out.

SPATULA

1. Forge a shoulder 1¾" (45 mm) from the end on the 1" (25 mm) face with half-face blows on the near side of the anvil and forge in a swell (a reverse taper out to the end). Neck in ¾" (20 mm) from the first shoulder on the ¼" (6 mm) faces.

2. Forge a trough down the center of the "proto" blade with a cross-peen hammer and gradually widen symmetrically. To avoid burning the edges, leave them thicker until finishing the spreading process.

3. Hammer out the peen marks of the spatula while holding the spatula in line with the horn of the anvil (the horn will act as a fuller to get some additional width in the blade).

HANDLE

1. Draw out the remainder of the handle to approximately 3/16" x ½" (5 mm x 12 mm).

2. Neck in the handle 1" (25 mm) from the blade to approximately ¼" (6 mm). Turn 90 degrees so the blade is upright and shoulder with a half-face blow.

3. Move up 1¼" (30 mm) and repeat the operation on the near edge of the anvil. Octagon and then round the isolated area to create a decorative "barrel."

4. With the cross-peen hammer, spread the shoulders on each side of the barrel. Do so at 45 degrees to create a little cusp, which greatly increases the strength of the transition points.
←

5. Offset ¾" (20 mm) from the top of the handle using half-face blows. This will become the hook.

6. Make a shallow fuller down the back of the handle from the barrel to the offset.

7. Forge the offset section to a point. Create a small curl in the direction of the offset, quench the curl, and then close the loop using the horn of the anvil. Kick the offset over 20 to 30 degrees before curling to create a rounded loop.

8. Correct any misalignment at this stage. A rawhide or wood mallet may be useful, as well as a twisting wrench. These operations can be done cold or hot, but not at a black heat as the metal is most brittle in that state.

9. Trim or grind any excess material from the spatula blade and cold hammer a chamfer into the spatula's edges. Refine with a file.

10. Use a low heat to set the angle of the blade relative to the handle as well as put an ergonomic curve into the handle. Wire brush and finish with a food-safe finish.

RAM'S HEAD MEAT FLIPPER

SUGGESTED MATERIALS

$^3/_8$" (10 mm) square stock run long

SUGGESTED TOOLS

- Hot cutter
- Flat chisel
- Scrolling tongs

TECHNIQUES

- Tapering
- Curling
- Punching

The ram's head is a traditional blacksmithing design element, especially for handles. While there are many different variations of the ram's head, this version is nice because it can be made with only a hot-cutting chisel and a center punch.

RAM'S HEAD

1. Forge a 3" (75 mm) square taper to a dull point. Keep the width of the taper at 3/8" (10 mm).

2. Use the chisel to make a series of closely spaced indents on the top and sides of the tapered section.

3. Cut the tapered section down the middle, work from the tip backward so you can see the line you are cutting. Clean the split on the corner of the anvil and then dress the edges of the "horns" with a file.

4. Round off 3" (75 mm) of the square stock starting 1½" (40 mm) from start of the horns. This will be the "neck" of the ram.

5. Hammer the horns over the edge of the anvil and then back onto the piece to create a "V" shape. Taper the end slightly, keeping the tip flat and blunt with backward blows as necessary. This "V" will become the face of the ram (the fourth step in the progression photo).

6. Hold the face (the part closer to the horns) at a half turn against the curved horn of the anvil and hit downward. Do this on both edges of the face to create a beveled surface that defines the nose and eyes.

←

7. Clamp the rounded neck into the vise. Use a center punch hammered at 45 degrees to the head to make the eyes. Use several blows as necessary. Make two more indents with the center punch for the nostrils. Finally, use a hot cutter under those holes to create the mouth. This step may take two or more heats.

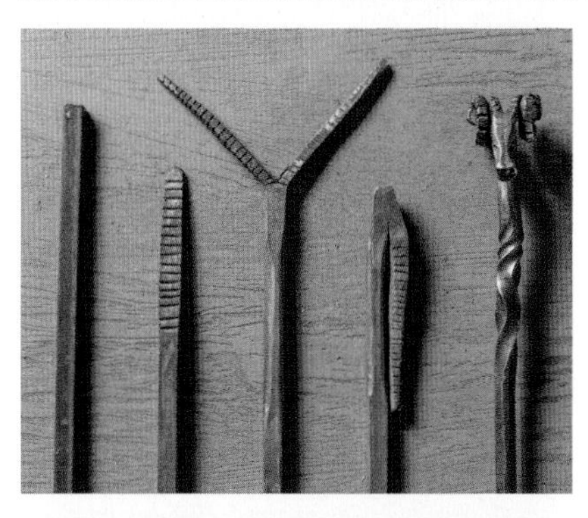

8. Get a good heat along the first 6" (150 mm), quench the horns and face, and then form the "head" and "neck" into a curve shaped like a question mark.

9. Use scrolling tongs to curve the horns into their final shape.

MEAT FLIPPER

1. Cut off the square stock at 8" (200 mm) from the head.

2. Taper this section to 20" (510 mm). Keep the first 10" (250 mm) from the head squared and the final 10" (250 mm) to the point rounded.

3. Quench. Heat the top portion of the shaft and head. Clamp the head into a vise and twist the first 2" (50 mm) of the shaft. A nice touch is to offset the twist by 45 degrees so that the shaft edge is aligned with the ram's head.

4. Quench. Heat the tip. Forge a half circle 1" (25 mm) in, then bend that half circle over the bevel of the anvil. Finish the bend by holding the shaft above and perpendicular to the anvil and hitting down on the half circle.

5. Clean the piece thoroughly, especially the tip, and coat with a food-safe oil.

PROJECTS

- Cooking Tripod and Spit
- Trammel Hook
- Plant Hanger
- Garden Tools
- Fire Brazier

CHAPTER 8: OUTDOOR

The products of the blacksmith are not limited to interior pieces. Whether for gardening around the house, camping with family and friends, or sitting around the fire pit on a cool summer evening, the projects in this chapter will help you bring your craft outside.

Unless you live in an extremely dry climate, moisture considerations will need to inform your choice of finishes for exterior pieces. An oil finish is a good option for the cooking tripod and trammel hook. I try to keep them indoors and out of the elements when not in use. For plant hangers, painting is a natural choice, but also consider leaving them bare and letting them develop a natural rust over time. Using garden tools quickly wears off finishes, but I wire brush the metal and use a good-quality lacquer at the end of each season to keep them looking fresh Finally, a high-heat paint—the type used to touch up BBQ grill covers—will help keep rust off your fire basket

COOKING TRIPOD AND SPIT

Not much beats cooking food over an open fire with tools you forged yourself in another fire not long before. This quick project gives multiple options for outdoor cooking needs. Install all three legs over the fire as a tripod or just the two "U" ends and put the third piece across the top as a spit. Use the same configuration without a fire and you have yourself an outdoor rack to hang pots and pans.

SUGGESTED MATERIALS

½" (12 mm) round bar, 3 pieces run long

SUGGESTED TOOLS

Medium bending forks

TECHNIQUES

Tapering

Curves

1. Forge a 3" (75 mm) loop on one end of one bar.

2. Forge a 2½" (65 mm) "U" on one end of the other two bars. Use medium scrolling forks to start a tight bend back onto the bar. Quench the "U," then close the bend on the horn of the anvil.

←

3. Cut all the pieces to the desired final lengths for the size of cooking tripod you would like. I've found a final height of about 36" (900 mm) is good for most fires, and I add 6" (150 mm) for hammering into the ground for a total length of 42" (105 mm).

4. Forge a taper in the straight ends, wire brush lightly, and then finish. Linseed oil makes a decent finish as long as the pieces are stored indoors when not in use.

COOKING TRAMMEL HOOK

SUGGESTED MATERIALS

- ½" (12 mm) round stock cut to 12" (300 mm)
- ¼" (6 mm) round bar cut to 18" (460 mm)

SUGGESTED TOOLS

- ⅜" (10 mm) punch

A trammel hook is a traditional cooking tool used to adjust the position of a pot placed over a fire. The hook in this project adjusts from about 13" (330 mm) to 20" (500 mm), but the design can be modified to fit larger or smaller dimensions. This trammel also has a loop incorporated into the hook to allow for adjusting the height using a stick or poker.

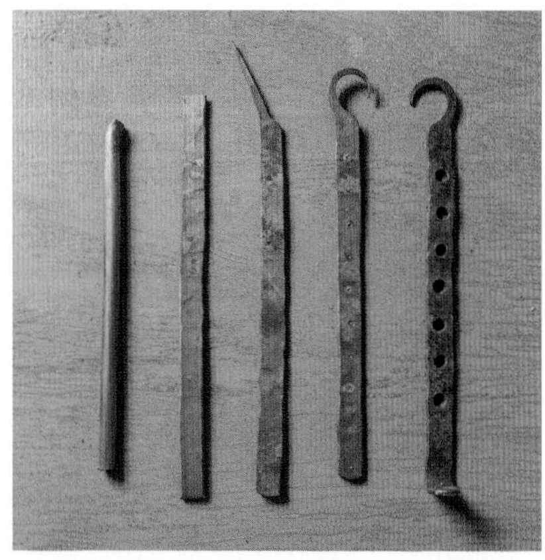

TECHNIQUES

Punching

Tapering

Bending

TOP HANGER

1. Flatten the ½" (12 mm) round stock to ¾" (20 mm) wide. Offset 1" (25 mm) of the flat stock using half-face blows and taper to 2½" to 3" (50 to 75 mm). Forge a 1½" (40 mm) hook from the taper in the same plane as the flat stock.

↓

2. From the other end of the flat stock, punch or drill ⅜" (10 mm) holes at ½" (12 mm), 3½" (90 mm), 4½" (115 mm), 5½" (140 mm), 6½" (165 mm), 7½" (190 mm), 8½" (215 mm), and 9½" (240 mm). Heat and bend the end without the hook at the 1" (25 mm) mark to a right angle.

BOTTOM HOOK

1. Create a 1½" (40 mm) loop at one end of the round bar with 1" (25 mm) of extra bar at the end. This can be done freehand or with bending forks.

2. Offset the loop so it will not hit the top hanger when assembled. Make the angle between the tab and the main shaft slightly sharper than 90 degrees, (about 60 to 70 degrees).

3. Make a 1½" (40 mm) "J" hook on the far end. Make sure the bends are loose enough to allow the piece to be attached to the top hanger through the top hanger's bottom hole. Wire brush and finish both pieces, then assemble.

PLANT HANGER

SUGGESTED MATERIALS	SUGGESTED TOOLS	TECHNIQUES
¼" x 1¼" (6 mm x 30 mm) flat stock cut at 24" (610 mm)	Bending forks	Tapering
¼" x ¾" (6 mm x 20 mm) flat stock cut at 20" (510 mm)	Riveting tool (optional)	Curving
		Upsetting (optional)
		Riveting (optional)

Whether for sale or personal use, it is useful to have different size hangers for plants, feeders, and signs. The benefit of this style hanger is it gives two different spots to hang items, each a different distance from the wall or post from which it is mounted. Create even more versatility by scaling the design larger and smaller using the same starting materials.

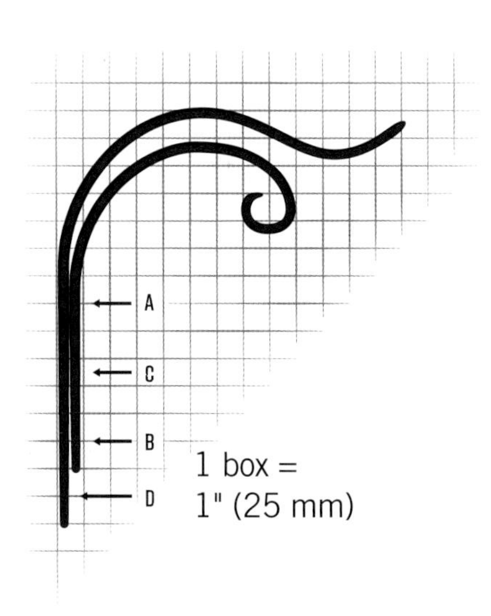

1. Use chalk and a steel square to layout the piece. If you have concrete floors, you can lay out projects in chalk directly on the floor. You can also use the surface of a steel welding table or a piece of Masonite board.

2. Forge an 11" (280 mm) taper into the ¼" x 1¼" (6 mm x 30 mm) flat stock and a 7" (180 mm) taper into the ¼" x ¾" (6 mm x 20 mm) flat stock.

3. For the outer hanger, begin creating the curve starting at the tip of the piece with gentle hammer blows over the horn of the anvil. Finish the curves using the anvil or bending forks.

4. Mark the straight end at 15" (380 mm) from the top corner on the template. **Simple option**: Cut off at the mark and dress the cut. Move onto next step. **More complex**: Cut off the piece diagonally at the mark. Get a tight heat on the end, clamp 1" (25 mm) from the end in a vise and upset. Bevel the upset edge on the anvil at the end of the heat. ←

5. For the inner hanger, create the initial curl on the anvil and finish the curve on the anvil or using bending forks, checking against the template often.

6. Mark the straight end at 13" (330 mm) and cut it based on one of the options above. If doing a diagonal cut, make it in the opposite direction as the outer hanger.

FINISHING: OPTION 1 (SIMPLE)

1. Drill ¼" (6 mm) holes in the spots indicated by the diagram, and use the mounting screws to attach the two hangers together when installing. Drill out the inner hanger holes first, then use those holes to check the marks for the outer hanger holes.

2. Wire brush and finish piece, making sure to use a finish appropriate for the outdoors.

OPTION 2 (MORE COMPLEX)

1. Drill out holes A, B, C, and D using a ¼" (6 mm) drill bit. Countersink holes A and B on the backside of the outer bracket using a wide countersink bit.

2. Rivet the pieces together using ¼" (6 mm) rivets, making the rivet flush on the backside of the brackets.

3. Wire brush and finish the piece, making sure to use a finish appropriate for the outdoors.

GARDEN TOOLS

SUGGESTED MATERIALS

- 14-gauge sheet metal (trowel, transplanter, hoe)
- ¼" x 1" (6 mm x 25 mm) flat stock run long (weeder)
- ¼" x 1½" (6 mm x 40 mm) flat stock, run long (cultivator)
- ⅜" (10 mm) square stock cut to 6" to 7" (150 to 175 mm) (shaft)
- $^{3}/_{16}$" (4 mm) round stock (rivets)
- Wooden handle blanks

SUGGESTED TOOLS

- Riveting tool
- Hot-cutting chisel
- Mallet

TECHNIQUES

- Cutting sheet metal
- Riveting
- Hot cutting
- Tapering
- Dishing

Blacksmiths were the original tool makers. Back in the day, a blacksmith-made shovel wasn't an artisanal option, it was the only option. These gardening tool projects are a fun way to carry on that tradition. The trowels and hoes are made of sheet metal riveted to the shafts while the cultivator and weeders are forged down from one piece of steel. For even longer-lasting tools, try using stainless steel or high carbon sheet metal for the working ends.

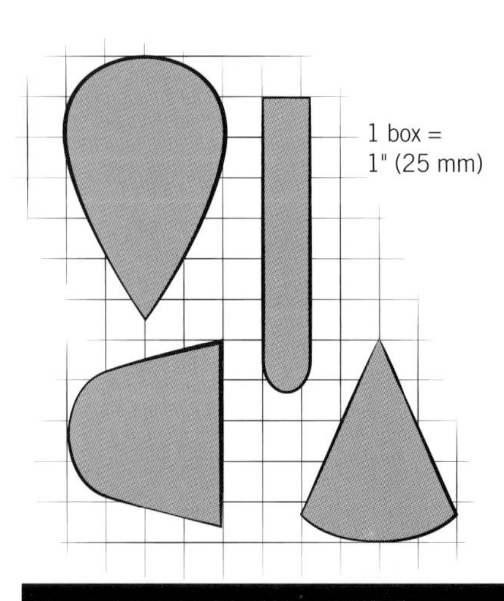

PART 1: WORKING ENDS

CULTIVATOR

1. Hot cut two evenly spaced 3" (75 mm) lines in the 1½" (40 mm) flat stock.

2. Forge each prong down to ¼" (6 mm) round with a tapered ends.

3. Forge the base material down to 3/8" (10 mm) square and 5" (125 mm) long, cut off the excess and forge a slight taper in the end to complete the tang.

4. Make sure the prongs are each the same length, get an even heat, and then clamp the base material in a vise and spread the prongs with small tongs and bend 90 degrees.

WEEDER

1. Hot cut down 1" (25 mm) down the center of the 1" (25 mm) flat stock.

2. Forge a slightly tapered point into each prong, maintaining the thickness of the stock.

3. Forge the base material down to 3/8" (10 mm) square and 7" (175 mm) long and cut off the excess. Forge a slight taper on the end for the tang (part of the metal that goes into the handle).

4. Get an even heat on the prongs and curve spread the ends slightly. Use a file to put a sharp bevel on the inside of the "V" created by the prongs.

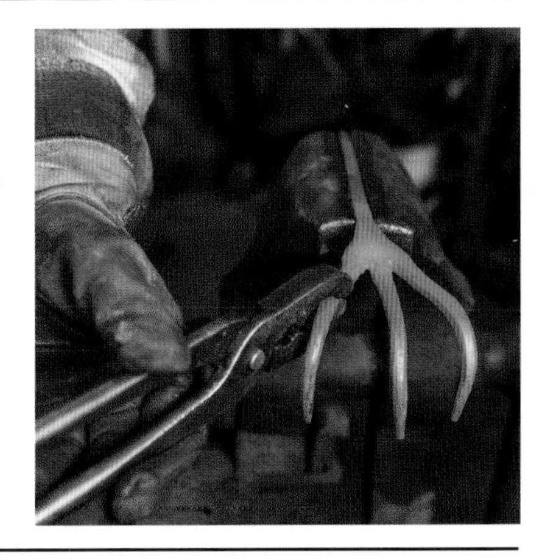

TROWELS

1. Cut the shape out of sheet metal, clean the edges, heat in the forge and dish using a cylindrical former or the step of the anvil.

2. Drill or punch two 3/16" (4 mm) holes for rivets about 1" (25 mm) and 2" (50 mm) into the piece and attach to the 3/8" (10 mm) square stock.

3. Forge a slight taper into the end of the tang.

4. For the traditional trowel, forge an offset into the shaft. Keep the shaft straight for the potting trowel.

HOES

1. Cut the shape out of sheet metal, clean the edges, heat in the forge, and create a slight curve in the piece over the edge of the anvil.

2. Drill or punch two 3/16" (4 mm) holes for rivets about 1" (25 mm) and 2" (50 mm) into the piece and attach to the 3/8" (10 mm) square stock.

3. Forge a slight taper into the end of the tang, then forge a curve slightly greater than 90 degrees into the shaft.

PART 2: ASSEMBLING THE HANDLES

Handles can be turned, carved by hand, or purchased pre-made. In either scenario, make sure the hole in the wooden handle is slightly smaller than the tang. The tang should fit tightly and need a mallet to be driven fully home. However, if the fit is too tight it will split the handle when being driven into place. If needed, drill out the hole until that tight, but not-too-tight, fit is achieved. Finish the handle and tools with a durable outdoor finish such as lacquer and refinish as needed.

SCOTTISH THISTLE FIRE BRAZIER

BY JIM WHITSON, THE BLAZING BLACKSMITH
PEEBLES, SCOTLAND, UK

SUGGESTED MATERIALS

- ½" (12 mm) round bar: 4 cut to 65" (1650 mm) (legs), 12 cut to 40" (1000 mm) (tops), and 1 cut to 32" (800 mm) (bottom rivet plate)
- ¼" x 1" (6 mm x 25 mm) flat bar cut to 70" (1775 mm)

SUGGESTED TOOLS

- Riveting tool
- Torch
- ⅜" (10 mm) round punch

TECHNIQUES

- Riveting
- Fish tails
- Using scrolling forks

Legend has it, an invading army was once trying to sneak up on an encampment of Scottish soldiers when one of the invaders stepped on a thistle and let out a scream, thus spoiling the attack. Ever since, the thistle has been the national emblem of Scotland. This fire brazier pays homage to that humble flower. It might not startle invading Norsemen, but it will certainly keep you warm on those cool summer nights.

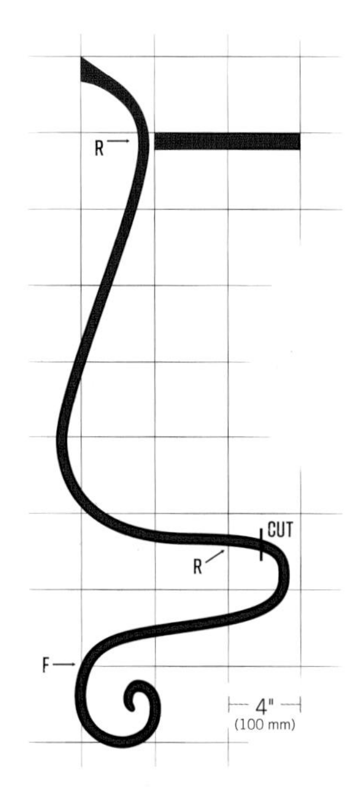

LEGS

1. Cut the end at 45 degrees, heat, and spread using a cross-peen hammer.

2. Curve the leg to match the template, starting from the top of the basket. When you get to the foot mark referenced on the diagram, run the remainder of the bar 7" (175 mm) long and cut off at that point. It helps to make a jig.

3. Forge a taper in the end and curl into a scroll until the length matches the template. Bend the scrolled end at a right angle to the plane of the rest of the basket.

→

4. Use a ball punch to create a depression in each of the rivet spots marked in the diagram. Then drill or punch 3/8" (10 mm) holes in each of the spots. Repeat all the steps for the other three legs.

TOP PIECES

1. Cut the end at 45 degrees, heat, and spread using a fuller hammer. Curve the leg to match the template, starting from the top of the basket. Cut off the remainder at the spot marked on the template and round the edges with a grinder or file.

2. Use a ball punch to create a depression in each of the rivet spots marked in the diagram. Then drill or punch 3/8" (10 mm) holes in each of the spots. Repeat for the other 11 top pieces.

RIVETING LARGER PIECES

Since this is a larger piece, riveting without a torch will be cumbersome in a coal forge, and not really possible using a gas forge, unless you can set rivets in one heat each. As an alternative, you can use bolts in place of rivets. Use the bolt heads on the outside of the piece and peen the heads to match the handmade aesthetic.

TOP AND BOTTOM LOOP

1. Draw an 18" (46 cm) diameter circle. Use a protractor or speed square to mark 22½-degree increments around the circle. Using medium scrolling tongs, gradually curve the flat stock to the circular form. This can be done cold. Be sure to do very small bends along the entire length of the stock, working the same length over and over to avoid kinks.

2. If you have access to a welder or are proficient in forge welding, cut off the excess stock on both ends and weld the loop together. Alternatively, cut one end with 2" (50 mm) of overlap in the circle. Heat that end, then hammer it flat against the beginning of the loop to create a tight overlap.

←

3. Mark and punch or drill each of the 16 rivet holes at 3/8" (10 mm).

4. Repeat with a 9" (230 mm) diameter circle for the bottom loop, using ½" (12 mm) round stock to make the circular form. Use a ball punch to enlarge each of the rivet spots before punching or drilling the holes. This circle is best made hot.

ASSEMBLY

1. Use 3/8" (10 mm) rivets to assemble the fire basket. Start with the four legs, then move onto the rest of the top pieces. Use bolts to assemble the entire basket initially, replacing each bolt with a rivet as the piece is assembled.

PROJECTS

- Steel Bracelet
- Copper Bracelet
- Celtic Brooch
- Spiral Pendant
- Dragon Pendant
- Earring Rack
- Jewelry Tree

CHAPTER 9: JEWELRY

Working on smaller forged projects demands and develops finer hammer-work ability. While the force required is less, the skill required is equal if not greater. A missed hit here and there on a 1" (25 mm) thick piece of stock can get cleaned up through the rest of the forging process. However, the same wiggle room simply doesn't exist for a piece that may only be 1" (25 mm) in length. Conversely, being able to hit precisely on larger pieces reduces the amount of wasted hits and effort. Regardless of what you usually make, mixing in some finer work from time to time will increase your abilities across all aspects of your craft.

This chapter includes a bracelet project in copper following one in steel. Switching between the scale of pieces as well as working with alternative materials can help stimulate new designs. In this case, both projects cover the same type of item, but the materials dictate a unique approach to each project. Working in softer metals also offers the weary smith a nice break from the much more stubborn steel from time to time.

STEEL BRACELET

MATERIALS	TOOLS	TECHNIQUES
$^{5}/_{16}$" (8 mm) round bar (length varies)	Small round tongs	Fullering
	Fullering tool	Tapering
	Additional depending on bracelet style	Additional depending on bracelet style

I started making bracelets as a project to work on with my girlfriend on one lazy Saturday afternoon. However, I quickly discovered how popular they were with customers and have been making them ever since. While the blacksmith will be required to make many pieces following precise designs, making jewelry is an aspect of metalworking where variation is a good thing; people like one-of-a-kind jewelry and will pay a premium for it. Once you've made a few bracelets, try combining styles, using different starting stocks, or even different materials such as bronze or copper (as with the next project).

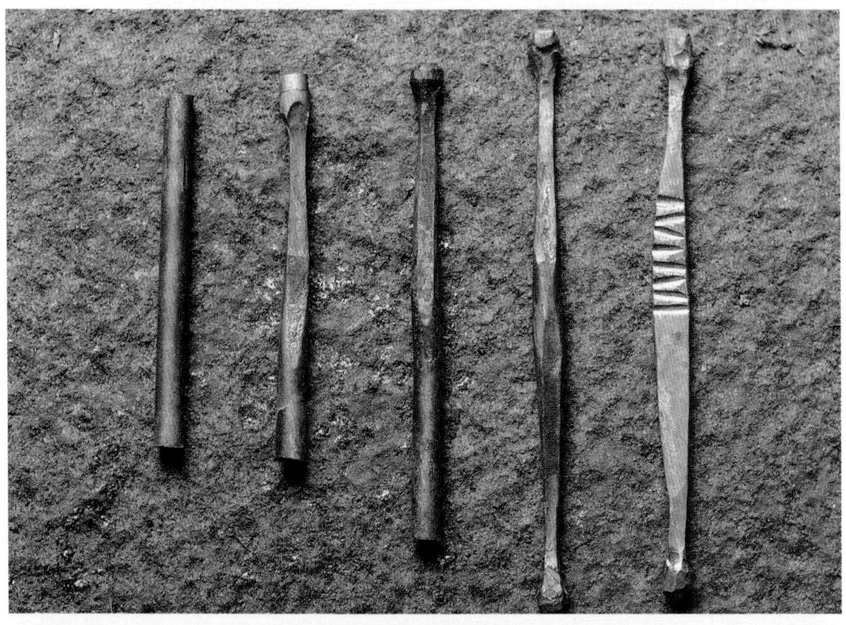

MAKING BRACELET BLANKS

1. Adjust starting lengths depending on your own style and size requirements. Starting with stock between 3½" and 4" (90 mm and 100 mm), drawn out to between 6" and 7" (150 to 180 mm) will cover most wrist sizes.

2. Use a fullering tool to isolate ⅜" (10 mm) at the end of the piece and taper the parent section of the stock slightly.

3. Create a ball on the end by rounding off the edge with angled blows while supporting the backside of the ball against the beveled edge of the anvil. As you hit down with the hammer on the front side of the ball, the beveled edge will help shape the backside. Finish rounding with light hits on the anvil's face. →

4. Continue tapering the parent stock, making a 2" to 3" (50 to 75 mm) taper depending on the desired final size of the bracelet. Repeat for the other end.

FINISHING

1. Get an even heat and curve the bracelet using the horn of the anvil. The final bracelet shape should be slightly oval rather than circular to fit on the wrist better. Work one end into the "C" shape, then the other end, adjusting as necessary. Be careful not to work the ends too cold as the stress can break off the ball ends.
→

2. Wire brush and finish according to desired final look. The brushed and waxed look highlights the unique surface that is created by hand working steel. It can also be accented with some brass brush burnishing or contrasted with a linseed oil finish on a portion of the bracelet. The skin's oils will also add some natural moisture protection when being worn.

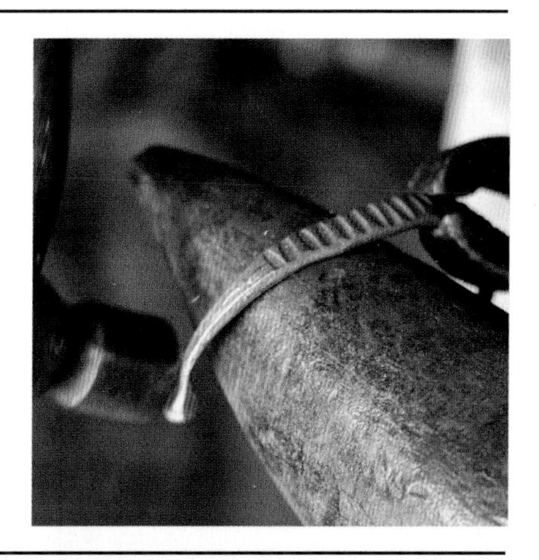

VARIATIONS

A: FLATTENING AND STAMPING

Get an even heat in the center of the piece, draw out and flatten until the total piece is the desired length: 6" (150 mm) for small wrists up to 7" (180 mm) for large wrists. Use chisels and punches to create different patterns on the flattened surface. This can sometimes be done cold.

B: TEXTURING

Draw out as described in part A, above. Use a fullering tool or ball peen to texture the surface. Using a fullering and ball punches instead of hammers, you can direct the blows with greater precision. Use a hold-down tool to free your hands while using the punches.
→

C: TWISTING

Draw out as in part A, above, but keep the stock square. Bevel the edges slightly. Twist the center of the piece using twisting forks. Experiment with different patterns of twists.

COPPER BRACELETS

MATERIALS

$^3/_8$" (10 mm) copper plumbing pipe

TOOLS

Clamps

Punches and chisels

Metal plate

TECHNIQUES

Texturing

Annealing

These bracelets made of repurposed plumbing pipe offer a enjoyable introduction to working with copper. Copper, in turn, makes for a natural addition to the modern blacksmith's repertoire since it can be worked with the same tools used to work steel. A note on the difference between copper and steel: Copper can be worked cold, but it will become brittle as it is worked. When it gets too brittle to continue forming, simply heat it to a "cherry red" with a torch or in the forge, then quench it or let it air cool. It will again be soft and ready to shape. This process of removing internal stress is called "annealing."

1. Cut the copper to length. 6½" (165 mm) will adjust to most wrists but go larger or smaller as needed. Flatten ½" (12 mm) of each end and round the edges with a file. Clamp the ends to a metal plate or large piece of flat stock secured on a workbench.

2. Texture the tubing (see variations at right). As you hit the tubing, surface dimension and textural variation is created. This is similar to chasing or embossing, the process of making a relief on a material's surface. Anneal as necessary when the copper becomes hard to work.

3. Anneal the bracelet again when texturing is complete, then wire brush the inside of the bracelet.

4. Use the horn of the anvil to curve the bracelet into a "C" shape. Begin by curving each end, then finish by curving the middle. Wire brush or polish the outside of the bracelet to finish.

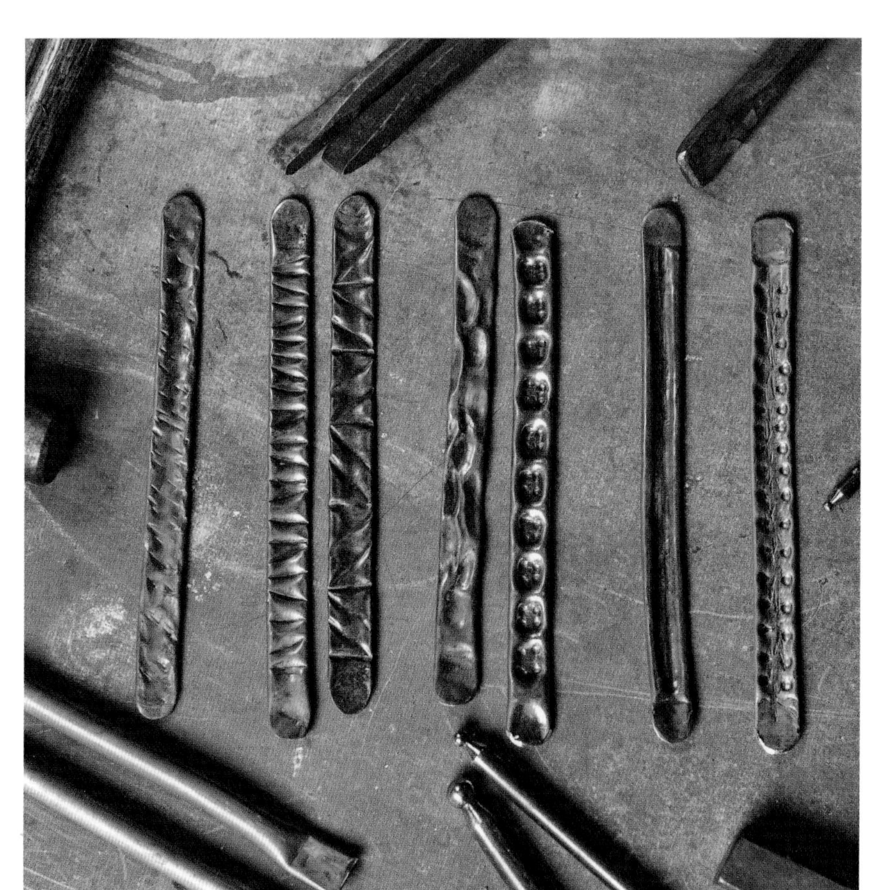

VARIATIONS

A. TEXTURING

Use the hammer face to slowly flatten the surface. Angle blows slightly and work the whole length of the piece to develop a lightly textured surface.

B. LINES

Flatten the tubing slightly and then use a dull chisel to create line patterns in the copper. Make them into regular patterns or irregular designs. Be careful not to hit too deeply on the edges with the chisel or risk cutting through the copper.

C. CRATERS

Use a small ball-peen hammer and various size ball punches to form craters along the surface. Once the initial indent is made, angle the ball punch to build up steeper ridges. This can also be done in regular or irregular variations.

D. RIDGES

Use a small ball punch or dull chisel to work along the centerline of the bracelet. Once this "valley" is established, continue working until the center line is flattened, then use angled blows from the chisel to build up the "ridges" on either side.

E. COMBINING TECHNIQUES

Begin by depressing the center line slightly as with ridges, above. Using a small ball punch, make regular indents along each side rail. After the initial depression, angle the blows toward the sides to build up the ridge. Go back along the center with a flat chisel and texture the copper along the center of the bracelet.

Copper will oxidize over time, losing a bit of its luster. It can also create some light greening on the skin of some folks. Neither of these things is harmful, but you can use a skin-safe sealer such as Everbrite to seal the piece and slow oxidation/greening.

CELTIC BROOCH

BY GUNVOR ANHOJ, CALNAN & ANHOJ
WICKLOW, IRELAND

MATERIALS
¼" (6 mm) round stock
½" (12 mm) round stock

TOOLS
Small and medium round tongs

TECHNIQUES
Tapering
Twisting
Scrolling
Chiseling

Celtic penannular (incomplete ring) brooches held a greater purpose in Celtic society than that of clothing fasteners. They served as secular and religious status symbols used by all genders, showing the rank, wealth, and importance of the wearer. Here are two traditional variations for you to try.

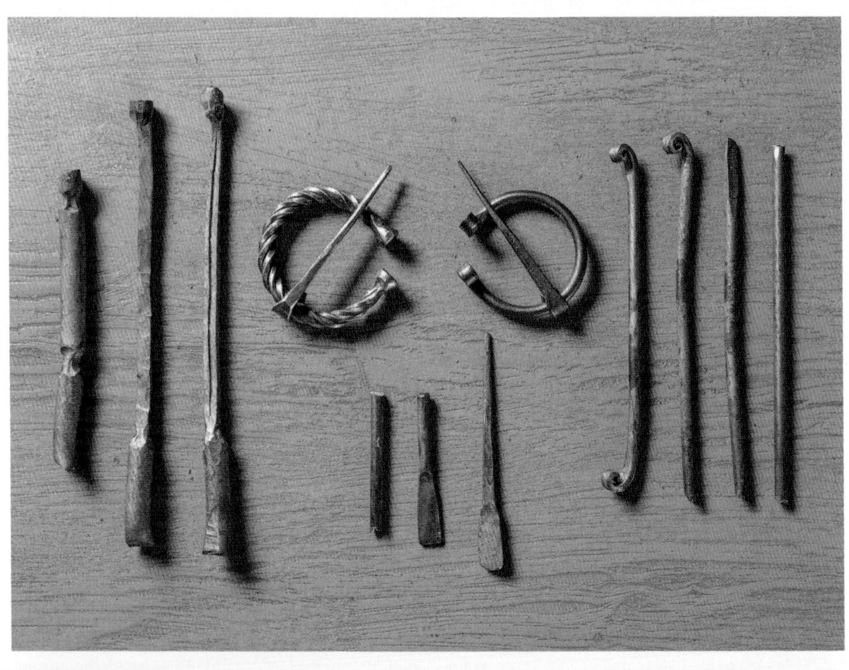

BROOCH WITH SCROLLED ENDS

This brooch uses the ¼" (6 mm) round stock.

1. Mark the piece at ¾" (20 mm) and create a 1¼" (30 mm) taper to a fine point, maintaining the width. Create a tight scroll back to the original mark, cut the piece at 6¾" (170 mm) from the scroll and repeat for the other end.

2. Shape the bar into an open ring. You can use a 2" (50 mm) bar or tube as a mandrel.

PIN

This pin uses the ¼" (6 mm) round stock.

1. Flatten 1¼" (30 mm) of an end of ¼" (6 mm) round stock run long.

2. Cut at 1¾" (45 mm) past the flattened section and taper to 3⅜" (85 mm). Start with a square taper and transition to a rounded taper at the point.

3. Pre-scroll the flattened end before closing around the brooch. Wire brush and finish.

BROOCH WITH "BUTTON HEAD" FINIALS

This brooch uses ½" (12 mm) round stock

1. Neck in the end of the bar using a spring fuller or similar. Neck in again 2½" (65 mm) from the end. Forge the material between the "necks" to an even ¼" (6 mm) square.

2. Chisel a line down the center of all sides. Get an even heat along the squared section and twist to desired pattern.
→

3. Cut the bar at the end of the second neck, and then clean the finals over the bevel of the anvil with backward blows while spinning the piece.

4. Shape the bar into an open ring, then follow the pin steps, above.

SPIRAL PENDANT

BY MARLEENA BARRAN, TAITAYA FORGE
CALBOURNE, ISLE OF WIGHT, UK

MATERIALS	TOOLS	TECHNIQUES
1/8" (3.5 mm) square stock cut to 2¾" (70 mm)	Small round nose pliers	Tapering
Leather cord	Small file or sandpaper	Curling

These small spiral pendants require care and precision to execute well. It is very easy for the piece to burn in the forge, so consider concentrating on one pendant at a time rather than forging several at once. Once you are comfortable with this spiral design, try out the dragon pendant, which builds on it.

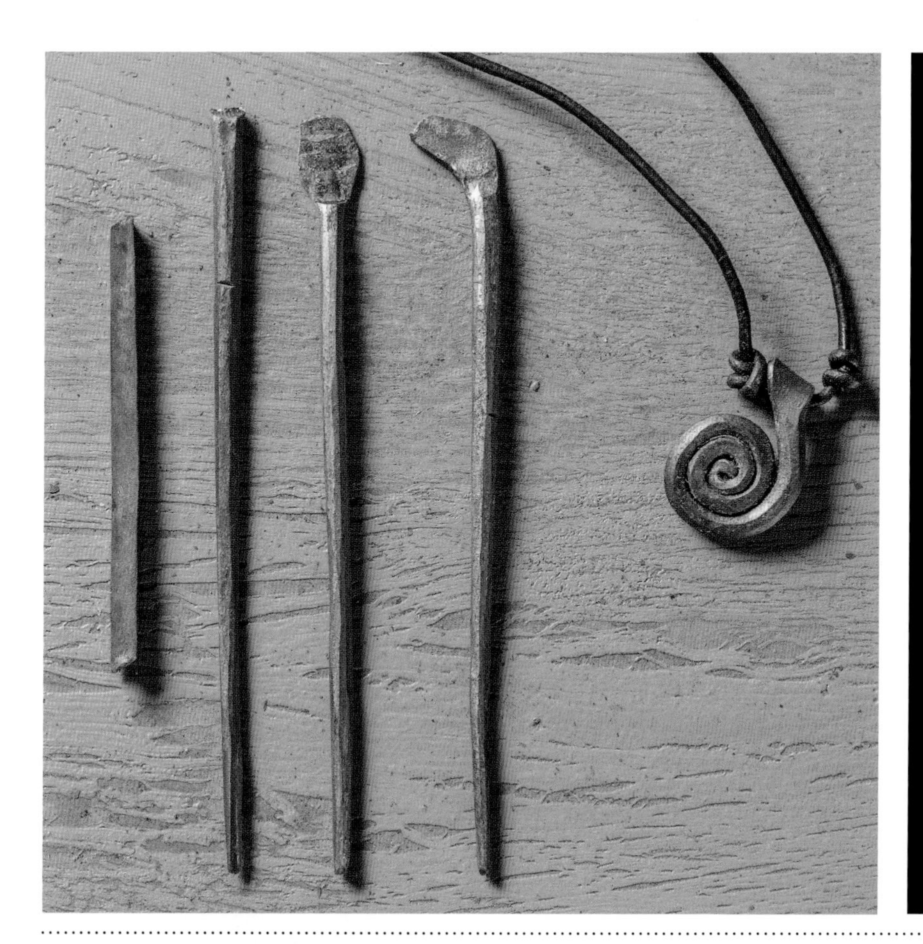

IN THE SHOP WITH TAITAYA FORGE

Marleena Barran got her start metalworking through an interest in Viking Age historical reenactment and love of jewelry from that time period. Much of her work today takes inspiration from that period. Marleena uses 99.8 percent pure iron for these and the dragon pendants. Pure iron is softer than mild steel, making it a suitable choice for delicate pieces. There are a few suppliers, such as Pureiron based in the UK, from which you can obtain some of your own.

1. Forge a point in one end, drawing the total length to 4¾" (120 mm) and roughly rounding the edges for a hand-textured look.

2. Flatten ⅜" (10 mm) of the end against the edge of the anvil.

3. Turn the piece 90 degrees and create a wedge-shaped bail (fastener) drawn out to ½" (12 mm). Angle the hammer at about 45 degrees to create the wedge. ↑

4. Spiral the tapered end. It is easiest to start cold and then heat the spiral.

5. Sand or file the edges of the bail so they are rounded and won't cut into the cord.

6. Bend the bail into shape using small round-nose pliers.

7. Wire brush and finish.

DRAGON PENDANT

BY MARLEENA BARRAN, TAITAYA FORGE
CALBOURNE, ISLE OF WIGHT, UK

MATERIALS	TOOLS	TECHNIQUES
3/16" (4 mm) square stock cut to 4 3/8" (110 mm)	Small round-nose pliers	Using chisels and punches
Leather cord	Needle files	Hot cutting
	Copper sheet	Curling
	Eye punch	Tool making

This dragon pendant builds on the spiral pendant project but incorporates several additional techniques to form the head of the dragon. As with the spiral pendant, take care when heating in the forge to avoid burning the stock.

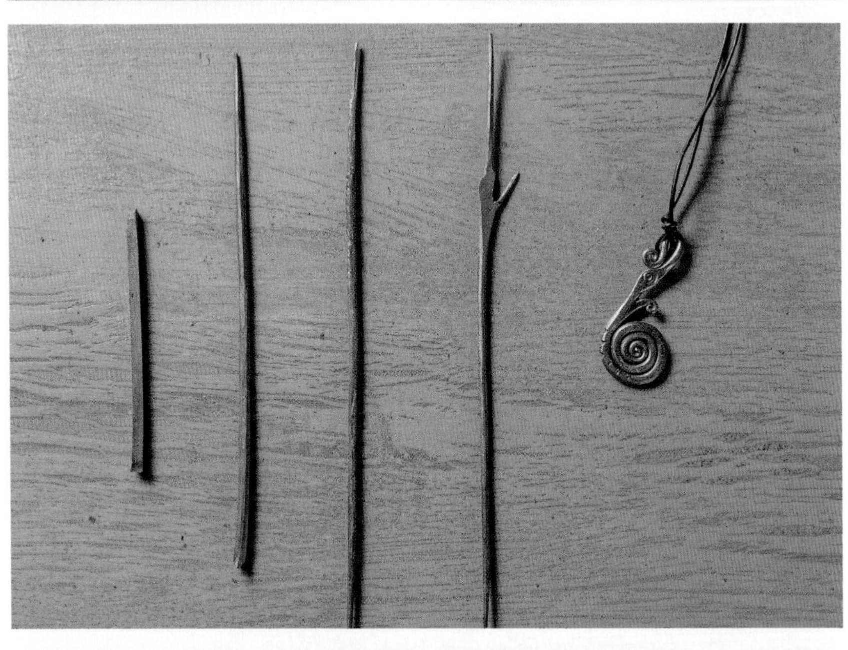

1. Forge a $8\frac{5}{8}$" (220 mm) round taper into the piece.

2. Forge a shoulder $\frac{7}{8}$" (20 mm) from the thick end and draw out to $2\frac{3}{8}$" (60 mm). Make the first $1\frac{1}{2}$" (40 mm) the same thickness and the last $\frac{7}{8}$" (20 mm) tapered to a point.

3. Make the transition point slightly off center and then flatten. Let the piece cool, then cold chisel the dragon's mouth. ↓

4. Fold the top lip back 90 degrees to isolate the bottom lip and forge the bottom lip into a tapered point.

5. Punch the eye. Use copper backing when punching the second eye to protect the first eye. See diagram below for the steps to make an eye punch.

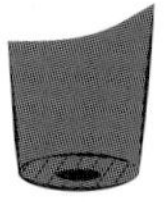

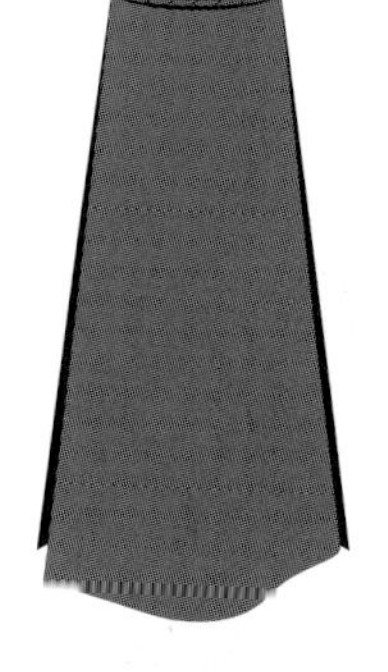

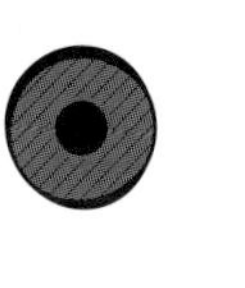

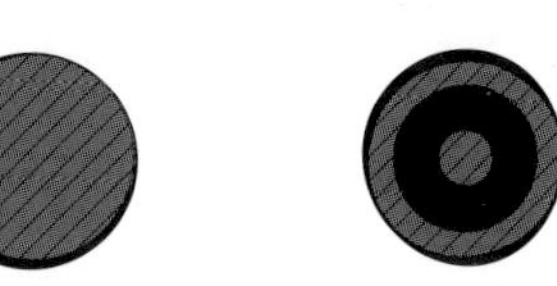

6. Clean the head and mouth with needle files. As an optional step, use needle files and a center punch to file grooves and a series of small craters down the neck.

7. Scroll the piece into the final form. For the top lip, begin by scrolling away from the shoulder until the length of the lip is $1\frac{1}{2}$" (40 mm) and then curve the lip back toward the shouldered section.

8. Wire brush and finish.

EARRING RACK

MATERIALS

- 1/8" x 1" (3 mm x 25 mm) flat stock: 2 cut to 14" (355 mm); 2 cut to 5" (125 mm)
- (Version A only) 2 cut to 7" (178 mm)
- (Version B only) 2 cut to 13" (325 mm)
- 3/16" (5 mm) round stock (rivets)
- Wire mesh cut to 6¾" x 13¾" (170 mm x 350 mm)

TOOLS

- Riveting tool
- Hot cutter
- Hand shears
- Clamps

TECHNIQUES

- Riveting
- Hot cutting (optional)
- Making leaves (optional)

This handmade earring rack is a beautiful way to display jewelry when it is not in use. For a classic look with a bit more fabrication than forging, try version A. For an organic twist, try version B, which incorporates some leaves and ivy.

1. Mark and drill out 3/16" (4 mm) holes centered at each end of the 5" (125 mm) and 14" (355 mm) flat stocks.

2. Mark and drill out additional holes at 2 1/12" (63 mm) on the 5" (125 mm) stock as well at 3½" (90 mm), 7" (178 mm), and 10½" (265 mm) on the 14" (355 mm) stock.

VERSION A

1. Temporarily assemble the frame and line up the 7" (178 mm) backing pieces under each end to line up and mark the rivet holes. Drill out the five holes on each 7" (178 mm) piece.

2. Assemble the frame again, this time insert the mesh behind the frame and in front of the 7" (178 mm) backing pieces. Clamp the entire frame together or use bolts through the rivet holes to temporarily hold the piece together during riveting.

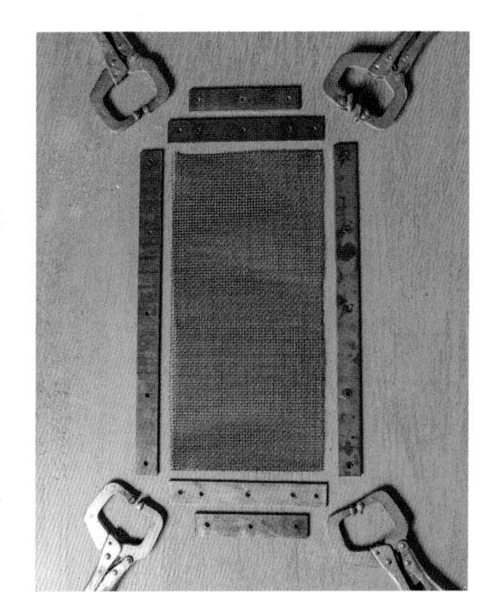

3. Rivet the piece, starting with the top and bottom of the frame. The three rivets in the center or each length are only used to secure the mesh backing.

4. Wire brush and finish the piece. If hanging, use a few small nails or screws through the mesh.

VERSION B

1. Hot cut or grind out a line 3" (75 mm) deep and ¼" (6 mm) from the side at each end of the 13" (325 mm) flat stock. Make the ¼" (6 mm) section at the opposite side of each end.

2. Fold the ¼" (6 mm) section back from the piece and forge a leaf with a 2½" to 3" (65 to 75 mm) taper out of the ¾" (20 mm) section.

3. Fold the completed leaf out of the way and fold the ¼" (6 mm) section back in line with the stock. Taper to a 3" (75 mm) round point.

4. Use scrolling tongs or pliers to form the ends into an organic shape.

←

5. Repeat on the other end.

6. Assemble and finish the piece following version A. Substitute these pieces for the 7" (178 mm) backing in version A.

JEWELRY TREE

MATERIALS

$^{3}/_{16}$" x 3" (4 mm x 75 mm) flat stock cut to 28" (710 mm)

TOOLS

- Hot cutter
- Half-cylinder dishing form

TECHNIQUES

- Hot cutting
- Dishing
- Texturing
- Curling

This jewelry tree highlights how much starting material can be transformed in the process of working a piece. The finished tree contains no stock that isn't altered in some way through the forging process. Organic designs such as these also leave room for variation. Feel free to experiment with cutting different size branches as well as how you spread them to create the final tree shape.

1. Mark the starting stock according to the diagram. Hot cut, angle grind, or cut with a band saw the lines that will become the branches in section A and the roots in section C. Forge an 18" (460 mm) taper out of the two 12" (300 mm) sections (section C).

2. Hammer section B with a sharp cross-peen hammer to give a bark texture.

3. Start forging the branches. Work each of the largest branches one section at a time, making a convenience bend to move the other sections out of the way while forging each branch.

←

4. Forge a round taper in each of the branch ends, then work back toward the main branches, rounding each section as you go. Repeat for the other two large branch sections.

5. Heat and dish section B into a semicylindrical shape. Make sure the bark is on the convex end. This can be done using a swage block, the step of the anvil, or the gap made by opening the jaws of a leg vise. Work from the edges in with a cross-peen hammer.

←

6. Fold section C perpendicular to the trunk. Forge gradual curves into the section, creating the base to support the tree with an organic, root-like form. Make sure it sits flat and that the tree comes up at a right angle from the base.

7. Heat and begin spreading the branches with scrolling tongs. Make sure that the end branches are arranged for jewelry to be able to hang off them. This can also be done with a torch.

8. Wire brush and finish the piece.

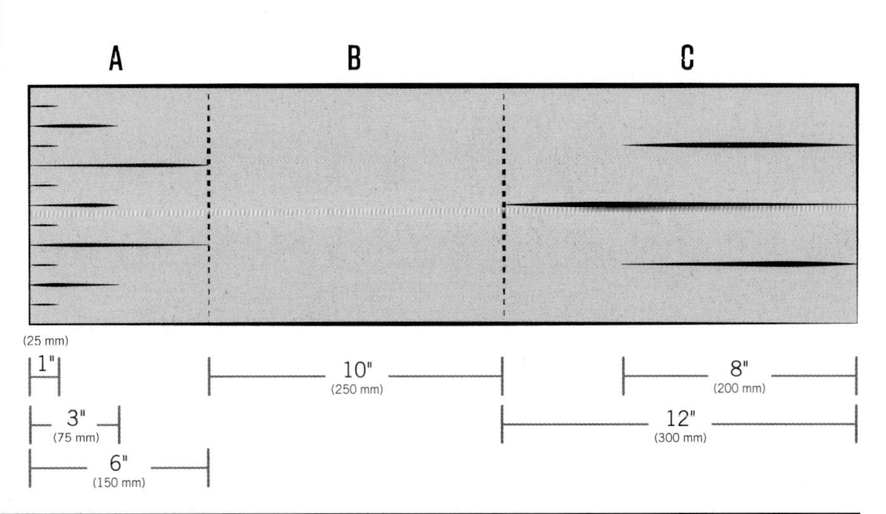

PROJECTS

CHAPTER 10: HARDWARE

Throughout blacksmithing history, the products of our craft have been used by folks in their everyday lives. Many of the projects throughout this book stay true to that everyday past, albeit with a modern twist. However, the final projects in this book—nails, hinges, and latches—tie in more directly to our blacksmithing heritage. These products enabled neighborhood homes to be built in a fashion that is still the model for modern houses. The process for making these projects hasn't changed much over the years, although the necessity of their use has, somewhat. I hope they serve as a reminder to the modern smith of our rich history of service to the communities in which we reside.

NAIL HEADER AND NAILS

SUGGESTED MATERIALS

- 3/4" (25 mm) round bar run long (another option is to use a tool steel)
- 1/4" (6 mm) round or square bar, run long

SUGGESTED TOOLS

- 1/8" (3 mm) square punch or needle file

TECHNIQUES

- Drawing out
- Punching
- Filing
- Heat treating (optional)

Making nails is a way to commune with blacksmithing history, practice your hammer technique, and—well, have nails to use! This nail header design is good because it doesn't require welding and creates headers for two different size nails in the same project. Nail headers may be made with mild steel, but a tool steel will last longer. There are several methods for making nails themselves. The following method is a good introductory approach. As you speed up, you may want to experiment with other methods.

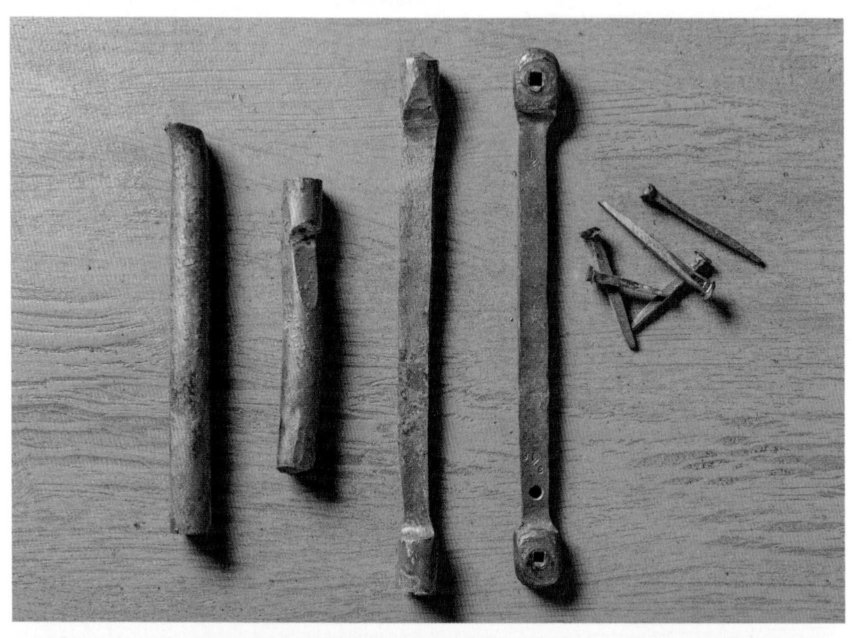

NAIL HEADER

1. Mark the round bar with a center punch at 1" (25 mm), 4" (100 mm), and 5" (125 mm).

2. Use half-face blows on the far side of the anvil to isolate 1" (25 mm) at the end of the bar. Work three sides so that this isolated boss is offset and raised. Do the same thing on the near side of the anvil with the 4" (100 mm) mark.

3. Draw out the center section to approximately 7" (175 mm), keeping the top face twice as wide as the side. Cut off the bar at the original 5" (125 mm) mark.

4. Round out the starting end and then hammer the top face into a dome shape. Repeat for the other end. If using tool steel, anneal the piece after this step by getting a good heat in the steel and letting it cool as slowly as possible.

5. Center punch each boss and drill out small, 1/8" (3 mm) pilot hole. From the underside, drill out a relief hole ½" (12 mm) in diameter and ½" (12 mm) deep. This will keep the tool from drawing too much heat from the nail being worked. Drill out another hole on the shaft near one of the bosses for hanging.

6. Heat the piece back up. Use the square punch from the underside of the boss to drift the pilot hole to the desired width. 3/16" (5 mm) and 1/4" (6 mm) are good medium sizes for general use.

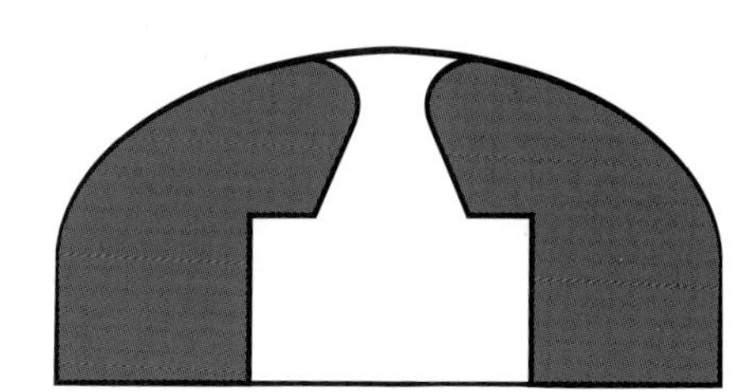

7. Use the punch again from the top side to create a slight taper in the hole. A cross section should look like an hourglass shape with a larger hourglass on the bottom. This will allow the header to grab the nails without them getting stuck.

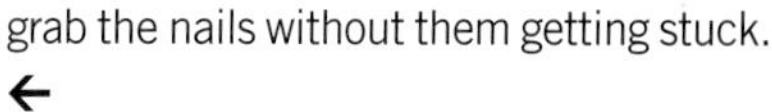

8. Repeat for the other boss, and clean the holes with needle files or emery paper so all the transitions are smooth. If using tool steel, harden and temper the piece at this time.

NAILS

These steps cover making 3/16" (4 mm) nails but can be followed for larger or smaller stock, substituting a starting material slightly larger than the nail header hole.

1. Taper the ¼" (6 mm) stock to the desired length. Turn the stock 90 degrees back and forth after each hit to minimize heat loss from the nail to the anvil and to create a shoulder for the nail head.

2. Center the shoulder at the end of the taper using a few half-face blows with the piece turned 180 degrees from the starting positions. Cut almost all the way through the nail head about 3/8" (10 mm) back from the taper.

3. If the piece has lost its heat, reheat at this point. If using a coal forge, bend the nail 90 degrees from the shaft to keep the tip out of the fire.

4. Place the nail in the header and twist off the shaft. Hit straight down (this is where your hammer control is tested) to upset the nail head. Use a few more blows to finish the head. Use angled blows around the header to make a rosette nail head or work straight down to make a flat head.

5. When finished, quench the header and the nail. The cooling will shrink the nail slightly, making it easier to knock out.

STRAP HINGE

SUGGESTED MATERIALS

- ⅛" x 1½" (3 mm x 40 mm) flat stock
- ¼" (6 mm) round stock

SUGGESTED TOOLS

- ¼" (6 mm) round drift
- Hot-cutting chisel
- Cold-cutting chisel
- Hacksaw
- File

TECHNIQUES

- Hot cutting
- Tapering
- Curling
- Riveting

Hinges have been used for more than 3,500 years. Their use even predates blacksmithing altogether. It is such an important piece of technology that the ancient Romans had a goddess of hinges—named Cardea—who protected the family from evil spirits. Whether you are also trying to keep out evil spirits or just that pesky draft, these hinges will do the trick. Depending on your application (e.g., door, cabinet), these project steps can be modified to make the leaves (ends) of the hinge larger or smaller. But the process for making the knuckles (joined portion) remains the same.

KNUCKLES

1. Bevel one end of the flat stock and begin to make a tight curl over the end of the anvil with the point of the bevel on the inside of the curve. Flip the piece over and continue curling with backward-facing blows.

2. When the knuckle is almost closed, insert the ¼" (6 mm) drift and use it as a mandrel (forming tool) to maintain the eye as you finish closing the knuckle. Take care to make sure the knuckle is kept perpendicular to the rest of the strap throughout this process. Repeat for the other leaf.

3. Divide each knuckle into thirds—in this case, ½" (12 mm) sections—and use a hacksaw to cut along each line. For the "male" leaf, use a cold-cutting chisel to finish removing the two outside eyes. For the "female" leaf, remove the inside eye. Use a file to even out the cuts match the leaves.

4. Make a rivet head on the ¼" (6 mm) round stock and then cut at 1¾" (45 mm). Insert the pin into the knuckles and move them about to make sure they move smoothly. If they are too tight, heat in the forge and then move them back and forth while hot.

LEAVES

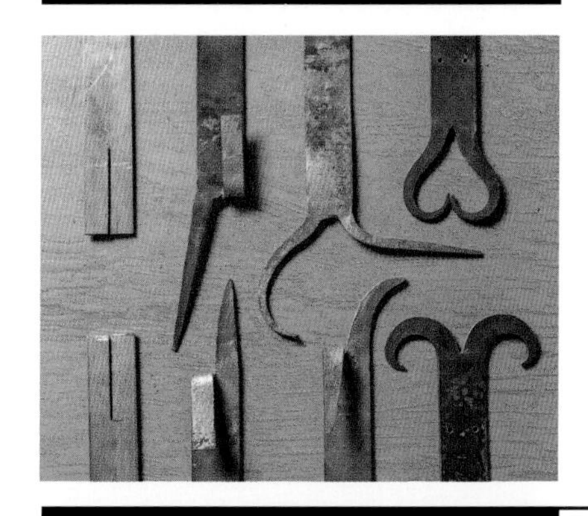

VARIATION 1: SHORT ENDS

1. Cut the leaf about 2" (50 mm) from where the knuckle ends. Round the ends and create a textured bevel with the peen of the hammer. Punch or drill three holes for mounting screws.

VARIATION 2: CURLED

1. Cut the leaf about 12" (300 mm) from the knuckle. This cut will determine the total length of your hinge, and you can adjust it based on the needs of your intended use (e.g., a door versus a cabinet).

2. Hot cut 2½" (65 mm) down the center of the end of the leaf and taper each section to 3" (75 mm).

3. Curl each taper over the horn of the anvil. As you make each curve, support the entire curve on the horn of the anvil and flatten often during the curving process to prevent kinks.

4. Punch or drill three or more holes for mounting screws along the length of the leaf.

VARIATION 3: HEART ENDS

1. Follow steps 1 and 2 for the curled leaf, but taper each section to 4½" to 5" (115 to 125 mm) instead of 3" (75 mm)

2. Open up the split ends into a "T" shape, then curl each end toward the split. Finish the heart shape by curling the middle of the curve over the horn of the anvil. It may take some tweaking to make the heart symmetrical.

3. Taper the leaf leading up to the heart slightly, quenching the heart before doing so. Punch or drill three or more holes for mounting screws along the length of the leaf.

SMALL LATCH

BY DAVID COURT, NORTHFIELD, NEW HAMPSHIRE, USA

SUGGESTED MATERIALS

- ½" (12 mm) square stock cut to 3½" (90 mm)
- ⅛" x ½" (3 mm x 12 mm) flat stock: 1 run long (lever/pull), 1 run long (70 mm) (retaining arm), and 1 cut to 4" to 5" (100 to 125 mm) (latch bar)
- ⅛" (3 mm) round stock (pin)

SUGGESTED TOOLS

- Small slot punch
- 3/16" x ¾" (5 mm x 20 mm) drift
- 3/16" (5 mm) spacer

TECHNIQUES

- Punching
- Drifting
- Spreading
- Tapering
- Curling
- Hot cutting

These next two projects show two ways to make traditional latch hardware. The mechanism for each is similar, but the starting stock used, as well as the scale of each, are quite different. As you head toward the final pages of this book and begin filling up shop notes with projects of your own design, remember a lesson from these last two projects: There are multiple ways to approach any design. Figure out what works best for you!

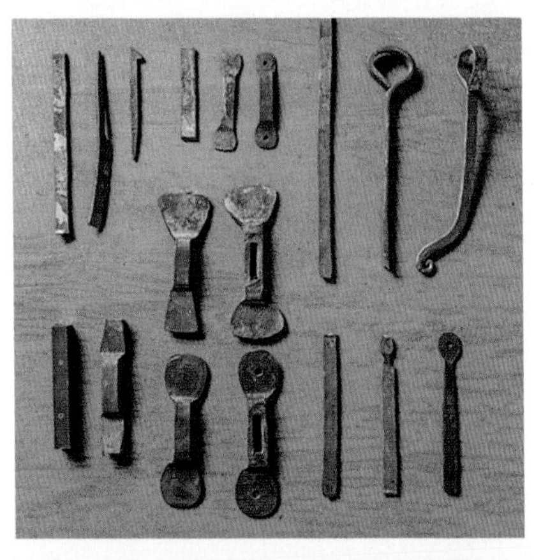

LATCH PLATE

Side set 1" (25 mm) from each end of the square stock, leaving 1½" (40 mm) of parent material in the middle.

OPTION 1:

Use a cross peen to spread out the ends into a bowtie shape.

OPTION 2:

Use the face of the hammer to make two rounded ends. Slot punch an 1/8" (3 mm) hole in the center of boss (center material) and drift the hole to 3/16" x ¾" (3 mm x 20 mm). Start the punch from the backside first, as this makes for a better-looking boss.

LEVER/PULL

Make an oval loop from approximately 4" (100 mm) of the flat stock. Cut the stock 5½" to 6" (137 to 150 mm) from the start of the loop, forge a 3" (75 mm) taper on the end of the flat stock, then forge a tight curl on the end followed by a gradual 90-degree bend in the opposite direction of the curl.

Make sure the curl is less than ¾" (20 mm) so it will fit through the latch plate. When adjusting the starting stock to fit your particular door width, make sure the curled end still outweighs the looped end based on the pivot point of assembly; otherwise the latch mechanism will not function properly.

ASSEMBLY

Make sure the thumb latch fits and moves freely in the handle. Enlarge the hole with a needle file as necessary. Drill through the handle and latch with a 1/8" (3 mm) drill bit and countersink the holes in the handle slightly. Cold rivet together with the 1/8" (3 mm) round stock.

LATCH BAR

Round the edges of one end of the 4" to 5" (100 to 125 mm) long flat stock and neck in ½" (12 mm) from the end. Forge a 1" to 2" (25 to 50 mm) taper to the neck. Flatten the end slightly and clean the edges into a circular shape. Drill or punch a hole in the center of the circle.

LATCH KEEPER

Offset and spread each end of the 2¾" (70 mm) long flat stock slightly using the face of the hammer. Flip the piece over and use a 3/16" (5 mm) spacer to forge a tight bend into the ends of the keeper. Drill or punch a hole at the center of each end.

RETAINING ARM

Forge an offset 1" (25 mm) from the end of a piece of 1/8" x ½" (3 mm x 12 mm) flat stock run long. Taper the offset to approximately 2½" (65 mm), then cut the parent material at a 45-degree angle to finish the catch. Clean the cut with a file.

INSTALLATION TIPS

When installing, attach the handle mechanism first (to the door), then install the latch bar and latch keeper according to the movement of the lever/pull. Finally, install the retaining arm so that the latch bar catches the latch bar when closing the door. The retaining arm is the only piece not installed on the door itself. Drill out a hole for the retaining arm nail slightly smaller than the nail itself; do not drive in the retaining arm too deep or the latch will be unable to catch on the arm.

HANDLE AND LATCH

BY JIM WHITSON, BLAZING BLACKSMITH
PEEBLES, SCOTLAND, UK

SUGGESTED MATERIALS

- ½" (12 mm) round stock cut to 6" (150 mm) (handle)
- ½" (12 mm) round stock run long (thumb latch)
- ½" (12 mm) round stock cut to 11½" (290 mm) (latch bar)
- ⅜" (10 mm) round stock cut to 3¾" (95 mm) (latch guide)
- ⅜" (10 mm) round stock run long (retaining arm)
- ⅛" (3 mm) round stock (pin)

SUGGESTED TOOLS

- Small rectangular punch
- $^{3}/_{16}$" x ¾" (5 mm x 20 mm) drift
- ¼" x 2" (6 mm x 50 mm) spacer

This traditional handle-and-latch project combines many techniques. However, what seems complicated as a whole becomes accessible when broken down into individual steps. Pay particular attention to making the handle mechanism—both the fit and the movement of the lever—to ensure proper functionality. Following the first part of this project, you can make a stand-alone door handle.

TECHNIQUES

- Leaves
- Hot cutting
- Tapering
- Punching

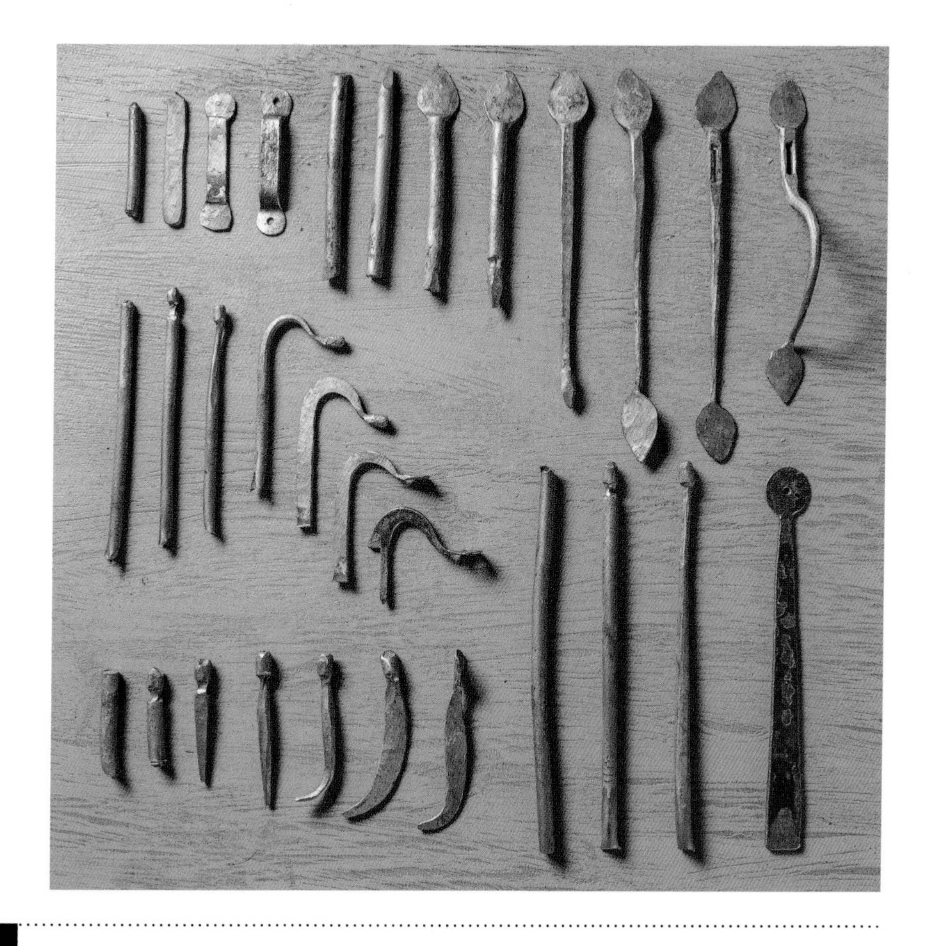

HANDLE

Forge the point for a leaf on one end of the round stock, but don't neck the leaf. Use half-face blows to create the leaf form on that end, then start forging a traditional leaf with a necked stem on the end of the stock. Square off the first 1" (25 mm) of the center stock leading to the first leaf (skip this for handle-only project), then forge a round taper down to the bottom leaf in remaining parent stock until the length of the center portion of the handle is 8½" (215 mm). Finish forging the second leaf at this time. Use a center punch to mark the dead center of the squared section, then use the punch and drift to make the hole for the thumb latch (skip this for handle-only project). Forge a curve into the handle and make sure the squared section and both lead ends are kept on the same plane.

THUMB LATCH

Round one end of the round stock and neck in about ½" (12 mm) from the end. Cut the piece off 2" (50 mm) from the neck, forge a round taper from the neck and then a smaller, reverse taper from the center of the section leading back to the neck for a total length of about 4¾" (120 mm). Curve the end of the tapered section and then flatten the whole tapered section. Flatten the rounded end in the opposite plane to the tapered section and then dish slightly with a ball peen-hammer. See photo on page 150.

ASSEMBLY

Make sure the thumb latch fits and moves freely in the handle. Enlarge the hole with a needle file as necessary. Drill through the handle and latch and countersink the holes in the handle slightly. Cold rivet together with the 1/8" (3 mm) round stock.

LATCH BAR

Round off one end of the round stock and neck in about 5/8" (16 mm) from the end. Forge a 9" (230 mm) round taper from the neck. Flatten the whole length of stock, and cut off the excess. Drill or punch a hole in the center of end circle.

LATCH GUIDE

Round each end of the round stock and then flatten the whole length of stock. Offset and further flatten the ends using half-face blows. Flip the piece over and use a 1/4" x 2" (6 mm x 50 mm) spacer to forge a tight bend into the ends of the guides. Drill or punch a hole at the center of each end of the latch guide.

RETAINING ARM

Round one end of the round stock and neck in about ½" (12 mm) from that end. Forge a 2" (50 mm) taper leading to the neck, then curve the taper into a half circle. Flatten all of the stock except the rounded end, then flatten the rounded end perpendicular to the end of the stock. Cut or grind away the top catch in the retaining arm to create space for the latch to rest as well as a built in nail for installation. Drill or punch a hole in the rounded end.

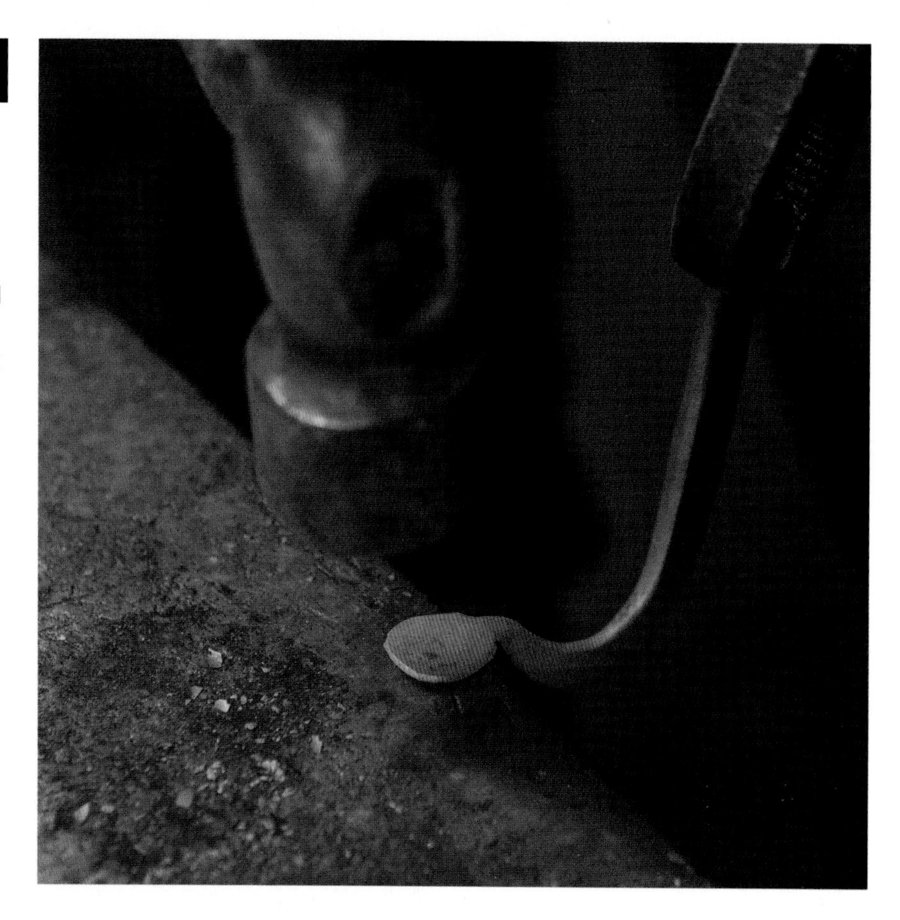

INSTALLATION TIPS

When installing, attach the handle mechanism first (to the door), then install the latch, latch catch (also called a "sneck"), and retaining arm according to the movement of the thumb latch. The retaining arm is the only piece not installed on the door itself. Drill out a hole for the retaining arm nail slightly smaller than the nail itself and do not drive in the retaining arm too deep or the latch will be unable to catch.

CONCLUSION

BLACKSMITHING TOMORROW

At the beginning of this book, we discussed a village blacksmith from the 1800s. Looking back to that time, we saw how technology has shifted but that there's still a demand for handmade goods. We also followed the blacksmith's evolving role within the community. As you wrap up the projects in this book, it's a good time to take a moment to think about the role you want to play in your community as a blacksmith, craftsperson, and artist.

Regardless of your skill and what you make, try not to underestimate the impacts you can create through your work. My first year blacksmithing, a customer ordered a personalized letter opener. I made the opener, gave it to the customer, and continued on with my other orders. Sometime later, the customer got in touch to tell me what that opener symbolized. She was quite close with her sister but had just moved across the country to attend university. The customer gave the opener to her sister along with a promise to write home often. That little letter opener—which took maybe 10 minutes to forge—became a reminder of their bond, strengthened every time her sister used it to open their latest correspondence. I've been fortunate to work on some great commissions in the years since, and I've received rave reviews, feedback, and publicity for those works. Yet, the bond between two siblings that my work helped reinforce still tops the list of rewarding experiences. It is also a reminder that we don't always know how our work will impact others, but we can bet it will.

In whatever direction you feel this craft is pulling you, whether as a weekend hobbyist or professional smith, I encourage you to continue making strides along that path. If our trade's history has taught us anything, it's that we can expect many changes ahead. There will be new pieces to make and new ways to share this great craft with family, friends, and neighbors. And it will be folks like you who determine just what direction blacksmithing goes in next.

GLOSSARY

ACETYLENE: Gas mixed with oxygen to create a high-temperature torch flame for shaping, cutting, and welding metal.

ALLOY: A substance with metallic properties made of multiple chemical elements.

ANGLE IRON: Term for steel sock containing right-angle cross sections.

ANNEAL: The process of softening metal through heating and slow cooling.

BALL PEEN: Hammer with a ball opposite the face, used for dishing, riveting, and other hammer actions.

BEVEL: To break the edges where two surfaces meet so they are not at right angles; also the term for this surface and the cutting edge of a tool.

BLACKSMITHING: The craft of heating and working iron and steel.

BOSS: A raised portion of the steel.

BOTTOM FULLER: Hardy tool for drawing and shaping metal.

BURNISH: To polish through rubbing, as with a wire brush.

CENTER PUNCH: Tool used to mark steel for drilling or other means.

CHISEL: Tool with a distinct blade for working steel in various ways such as cutting and indenting.

COLD CHISEL: Chisel with a shallower blade for cutting steel while cold.

COLLAR: To use a strip or band of metal to join two or more pieces of stock.

COUNTERSINK: Enlarging or beveling a hole so that a screw or rivet can sit flush with the hole.

CROSS-PEEN HAMMER: Hammer with a dull blade-like back (peen) used to spread and shape metal.

DRAW: To draw out, or draw to, is to elongate the metal being worked to a specific length or shape (e.g., draw to a point).

DRIFT: Tool for enlarging holes.

EYE: The part of hammer or other handled tool in which the handle is attached.

FACE: The main surface of the hammer.

FLUSH: Even with the surface.

FORGE: Container used to heat metal (i.e., put the metal in the forge); process of shaping hot metal (i.e., forge a short taper); name for a blacksmith shop (e.g., John's forge is located down the road).

FULL FACE: Blows of the hammer fully over the anvil.

FULLER: Tool with various configurations used to neck or shoulder iron with great accuracy.

GALVANIZED: Metal coated in zinc for rust protection; produces extremely harmful gases when heated.

GAUGE: Size measurement used primarily for thinner materials and rods; smaller gauge numbers equate to larger dimensions.

GRIT: The abrasive bits used to sand and polish surfaces; finer grits indicate larger numbers.

HALF FACE: Blows of the hammer aimed with half the face over the anvil and half off the anvil, used for actions such as off-setting.

HARDENING: The process in heat-treating used to maximizing hardness/brittleness in steel.

HARDY HOLE: Square hole on the anvil used to hold various tools.

HARDY CHISEL: Tool used in a hardy hole with a sharp edge for cutting metal.

HARDY TOOL: Any tool used in the hardy hole.

HEAD: To create a thickened end, such as on a rivet or nail.

HEADER: Tool used to head metals.

HEAT: The period of time a piece of metal from the forge can be worked before it becomes too cold.

HEAT TREATING: The process of regulating hardness in metal by controlled heating and cooling.

HIGH-CARBON STEEL: Also known as tool steel, steels with a higher carbon content allowing them to be hardened and tempered for use as tools.

HOLD-DOWN: Any of various tools used to keep metal in place and free the hands for other actions while it is being worked.

HORN: Rounded protrusion of the anvil, used primarily for making curves, also known as the "bick."

HOT-CUT CHISEL: Chisel with an angled blade for cutting hot steel.

HOT SETT: Handled tool for hot cutting metal.

IRON: A metallic element that is the primary component of all steels; the term is often used interchangeably with "steel."

JAWS: Gripping surface of tongs.

JIG: A device used to guide tools, especially in making repetitive forms.

LEG VISE: Vise with a large metal leg that can be anchored for additional stability.

MALLEABLE: The ability to be shaped.

MANDRIL: Device used for shaping metal pieces around, such as rings or the eyes of tools, often conical in shape.

MILD STEEL: A type of steel with a lower carbon content used by modern blacksmiths for general work because it has a good balance of strength and workability.

MONKEY TOOL: A hollow tool for shaping shoulders of a tenon.

NECK: To offset or indent all around a piece of stock. Compare with "Shoulder."

OXIDIZE: To combine with oxygen; causes scale to develop on the surface of steel as it is worked.

PATINA: Colored oxidation of metal surface, often used deliberately as a finishing process.

PEEN: The end of the hammer opposite the face, often of a specific variety (e.g., ball peen, cross peen).

PICKLING: Removing surface oxides through chemical processes.

PITTING: Cavities in metal caused by burning, oxidation, or corrosion over time.

PLIERS: Small griping tools.

PRITCHEL: The round hole on an anvil face, named for a specific tool called a "pritchel" used by farriers to punch nail holes in horse shoes.

PUNCH: Used to describe both the tool and the process of stamping a depression in, or creating a hole in metal.

QUENCH: To rapidly cool through immersion in liquid.

RASP: A coarse file used to remove large amounts of material quickly.

RIVET: To join with a bar through the steel with flattened ends; also the term for the hardware used in that process.

RIVETING TOOL: A tool with a hollow end for supporting or forming a rivet head.

RUN LONG: Keeping stock long during forging for handling without tongs and cutting to length at a later step.

SCALE: The bluish-black surface that develops on steel when it is heated due to the oxidation process.

SHOULDER: To offset or indent on one or two sides of stock. Compare with "Neck."

SMITH: One who works metals.

SMITHY: A shop where metal is worked.

SPRING SWAGE: Top and bottom swages linked with a springy handle.

STAINLESS STEEL: Steel alloyed with additional chemical elements to resist corrosion.

STEEL: Term describing a family of alloys made up primarily of iron and carbon.

STOCK: Term for metal being handled (e.g., square stock).

SWAGE: The process of, and any of various tools used to shape metal in specific ways. Some swage tools fit in the hardy hole of an anvil, others are handled and used with a power hammer.

SWAGE BLOCK: Large piece of metal with various cavities used to shape metal in specific ways.

TANG: End of a metal tool, which is attached to a handle.

TEMPER: The final step in heat treating, used to regulate the hardness/brittleness of metal.

TINES: Another term for prongs, or protrusions, of a piece.

TONGS: Tools used to grip steel.

TOOL STEEL: Another term for high-carbon steel.

UPSET: To thicken a portion of material (as opposed to elongating by drawing out).

VISE: Device for holding material securely while being worked.

WELD: To fuse two metals together through heating beyond their melting point.

WROUGHT IRON: A material once mass-produced and used by the blacksmith, no longer in wide production; mild steel is primarily used in its place.

ACKNOWLEDGMENTS

While writing is a solitary pursuit, writing a book is a group effort. I would like to thank the entire group at Quarto, whose combined experience and effort made the entire process not only possible but enjoyable.

Thanks especially to the *Everyday Blacksmith* team: Jonathan Simcosky, Meredith Quinn, David Martinell, and Steve Roth.

Thanks to Matt Kiedaisch of Outsider Media and Mark Coletti of Moose Art Designs for going above and beyond with the photography and illustrations; to Amy Paradysz for copyediting my jumbled scrawl (all mistakes are hers, not mine); and to Paul Burgess of Burge Agency for putting everyone's creative effort together into the book you see before you.

Not much beats working with great artists, except working with great artists who also happen to be great people. It was a pleasure working with all of you.

Many blacksmiths added their wisdom to the alchemy that made this project come to life. Locally, thanks to the contributors David Court, Bob Menard, Dereck Glaser, and Nicholas Downing for talking shop and providing much needed input and advice. And to all of the project contributors across the globe, my sincerest thanks and gratitude goes out to you. One of the most rewarding parts of working on this book was getting to meet and learn from all of you. Thanks for taking time out from running your shops to share your projects and passion for this great craft with both the readers and myself.

I've been blessed with many mentors over the years who helped me on the path to writing this book. Thanks to my father, Stuart, for teaching me how to wield a hammer and my mother, Joan, for teaching me how to wield a paintbrush. To Trevor Youngberg, thanks for teaching me how to combine aesthetics and craftsmanship. Thank you to Jim Whitson, for passing on this wonderful craft. To Peter Martin, Ryan D'agostino, and the folks at *Popular Mechanics*, thanks for giving me my first platform to share my perspective. Thanks also to Jennifer Plowden and Jessica Sargent for teaching me the skills to take this project on. To Ansley Newton and David Pollack, thank you for making Maine a place where I could grow my shop.

And to all my friends and family, I apologize for my endless chatter about "the book" and its progress. Thanks for listening.

Finally, to you the reader, thanks for continuing to nurture your passion for blacksmithing. It is folks like you who will keep this ancient tradition alive and moving it in exciting directions yet to be explored.

I look forward to seeing what becomes of your pursuits.

ABOUT THE CONTRIBUTORS

Gunvor Anhøj saw the importance and potential of iron for making and repairing everyday tools while working on a traditional horse-ploughed farm in Norway. More than 20 years since then, Gunvor has worked with metal in many capacities but enjoys sculpture most. She operates Calnan & Anhøj based in County Wicklow, Ireland. www.calnan-anhoj.ie

Marleena Barran loves the magic that comes from working with fire and shaping glowing-hot metal. Her favorite tools are tongs, because if you have the right tongs for the piece you're working, everything else becomes easier. Accordingly, her advice to new smiths is to learn to make your own tools! She runs Taitaya Forge out of the Isle of Wight in the United Kingdom. ww.taitaya.co.uk

Finín Liam Christi was inspired begin metalworking by his late grandfather who was blacksmith. He has been blacksmithing for 40 years and enjoys carrying on the art of blacksmithing through working with traditional tools and techniques. He is based in County Wexford, Ireland.

David Court has learned a thing or two about blacksmithing during his 50 years in the craft. But David is a born tinkerer and always studying new aspects of metalworking. In his free time, he has lately been exploring a traditional Japanese form of working sheet metal called *Uchidashi*. David is based in New Hampshire.

Chris Danby enjoyed working with his hands and fell in love with blacksmithing after enrolling on a training course many years ago. He and his wife established Coach House Forge based in Cheshire, England, where they share the many essential tasks involved in running a successful business. www.coachhouseforge.com

Nicholas K. Downing built his first forge from a tire rim, a shop vacuum, and a piece of a pottery kiln after taking a metalworking class in middle school. From those roots, he's grown into a well-respected blacksmith, jeweler, and instructor in New England. Nicholas is based in Maine. www.downingarts.com

David Gagne has a do-it-yourself ethic and a punk rock attitude. A born maker, he started Elm City Vintage in 2008 to offer unique handmade pieces at affordable prices. He is based in Connecticut. www.etsy.com/shop/elmcityvintage

Dereck Glaser has been dedicated to the design, forging, and fabrication of unique functional metalwork and metal art since 1985. After finishing a degree in Industrial Art Education, Dereck began a self-directed journey-manship throughout the eastern half of the United States. He eventually settled in Maine, where he has operated DG Forge for more than 20 years. www.dereckglaser.com

Matt Jenkins honed his skills and learned traditional blacksmith techniques while studying with master smiths around the world for 25 years. Between hammering on custom projects in his shop, Coverdale Forge, north of Winnipeg, Manitoba, Matt leads workshops and demonstrates the ancient craft of blacksmithing across North America. www.cloverdaleforge.com

Bill Kirkley became interested in blacksmithing after years of doing machine work in his free time. Accordingly, he likes combining these passions to make custom tooling and small machines. He splits his time between Maine and South Carolina.

Bob Menard was first exposed to metalworking as part of an experimental education program in Vermont where, for an early American history class, his class built a traditional log cabin complete with blacksmith-made hardware. For more than 40 years since then, Bob has been helping grow the craft in New England and beyond. He runs Ball and Chain Forge out of Maine. www.ballandchainforge.com

Chris Moore trained with some of the UK's top blacksmiths, working on projects for Buckingham Palace and St. Paul's Cathedral, among others. Back home in New Zealand in 2004, he founded Artistic Ironwork. www.artisticironwork.co.nz

Caitlin Morris took her first blacksmithing course in 2009 and fell in love with the craft immediately. After many years learning from and working with top blacksmiths from across the continent, in 2015 she began Ms. Caitlin's School of Blacksmithing, where everyone has a chance to learn blacksmithing, regardless of whether they "look like a blacksmith." She is based in Maryland. www.mscaitlinsschool.com

William Pinder has had a lifelong fascination with industrial history that eventually led him to set up his own blacksmith shop, where he loves making parts for vintage machinery using his British-built, belt-driven, 690-pound Goliath power hammer. He runs W. A. Pinder Ironworks, Victorian Forge and Millwrights out of Dorchester, England.

Kerry Rhodes grew up with an art teacher mother and photographer father, combines an artistic bent with a particular interest in the beauty of natural forms. He started Forged Creations in 1990 and is based in Delaware. www.forgedcreations.com

ABOUT THE TEAM

Photo Credit: Karla Bernstein

AUTHOR

Nicholas Wicks is a blacksmith and writer based in Maine. His family has been metalworking for five generations, starting with Nick's great-great-grandfather who worked on the Statue of Liberty. After apprenticing under Jim Whitson in Scotland, Nick started Wicks Forge in his grandfather's garage using homemade tools and his great-grandfather's anvil. Wicks Forge quickly expanded as audiences responded to Nick's products and philosophy—everyday pieces that combined a formal background in blacksmithing with an accessible DIY mentality. Nick started helping others join the Maker Movement as a contributing editor for *Popular Mechanics*, where he writes how-to articles and reviews about which work pants are least likely to catch on fire. Wicks Forge has been featured in numerous galleries and publications including *Yankee* magazine and *Connecticut Magazine*. Nick is a member of the New England Blacksmiths Guild and the Artist Blacksmith Association of North America. www.wicksforge.com

TECHNICAL EXPERT

Jim Whitson is an artist blacksmith based outside of Edinburgh, Scotland. He has more than 25 years of experience making one-of-a-kind metalwork, from flowing staircases of ivy, to classic gates, and publicly commissioned sculptures. Jim's roots in Scotland link him to an unbroken tradition of blacksmithing dating back centuries. His technical expertise stems from this tradition combined with a mastery of modern tools and techniques. Jim is a well-respected member the British Artist Blacksmith Association, and his award-winning designs have been featured across Europe. www.blazingblacksmith.co.uk

ILLUSTRATOR

Mark "Moose" Coletti is an artist specializing in graphic design and digital illustration. Currently based out of Boston, his work ranges from magazine layout and publication to custom t-shirt design and everything in between. Nick and Moose shared formative years at McGill University and formed an even deeper bond on a memorable road trip across America. Their collaboration for *Everyday Blacksmith* is one built on friendship and their shared love to create amazing things, every day. www.markcoletti.com.

PHOTOGRAPHER

Matt Kiedaisch is a visual artist specializing in photography and film. His interest in creating powerful images is rooted early experiences exploring art and architectural exhibits as a child. He now lives in Vermont with his family and travels widely to work with clients and friends. www.thinkoutsider.com

INDEX